Harmony with Nature: Exploring Environmental Engineering Solutions

In a world faced with increasing environmental challenges, the field of environmental engineering emerges as a guiding force in seeking sustainable solutions for a harmonious coexistence between humanity and the natural world. "Harmony with Nature: Exploring Environmental Engineering Solutions" is a comprehensive exploration of the vital role environmental engineering plays in tackling pressing environmental issues and paving the way for a more sustainable future.

This book delves into the dynamic and interdisciplinary field of environmental engineering, offering a deep dive into the principles, practices, and innovative approaches employed to address a range of environmental concerns. From the preservation of natural resources to the mitigation of pollution, from the design of eco-friendly infrastructure to the promotion of renewable energy, this book serves as a compass for understanding the multifaceted challenges faced by our planet and the engineering solutions that can help restore harmony.

"Harmony with Nature" illuminates the importance of adopting an integrated approach that considers the intricate interconnections between human activities and the environment. It explores the intricate web of environmental systems, highlighting the delicate balance that must be maintained to ensure the long-term sustainability of our planet.

Through engaging narratives and case studies, this book showcases the innovative techniques and technologies used by environmental engineers to combat pollution, protect ecosystems, and promote a more sustainable future. It delves into topics such as water resource management, waste management, air quality control, sustainable urban development, and climate change adaptation, providing readers with a comprehensive understanding of the challenges and solutions in the realm of environmental engineering.

Moreover, "Harmony with Nature" emphasizes the importance of collaboration, policy implementation, and public engagement in achieving environmental goals. It examines the role of governments, organizations, and individuals in driving change and adopting sustainable practices in both developed and developing regions.

Whether you are an environmental engineering professional, a student pursuing a career in the field, or simply an individual interested in understanding and mitigating the environmental challenges we face, "Harmony with Nature" offers valuable insights and practical guidance. It serves as a source of inspiration and a call to action, encouraging readers to play an active role in promoting a harmonious relationship with nature and ensuring the well-being of future generations.

Join us on this enlightening journey as we explore the realm of environmental engineering and discover the pathways to a sustainable and harmonious future. Together, let us embrace the principles of environmental stewardship and work towards a world where humanity and nature can thrive in perfect balance.

I. Introduction

- The urgency of environmental challenges

- The role of environmental engineering in finding sustainable solutions
- The significance of achieving harmony with nature

II. Understanding Environmental Engineering

- Definition and scope of environmental engineering
- Interdisciplinary nature of the field
- The importance of systems thinking and holistic approaches

III. Environmental Systems and Challenges

- Overview of key environmental systems (e.g., water, air, land)
- Examination of global environmental challenges (e.g., climate change, pollution, resource depletion)
- Impacts of human activities on the environment

IV. Principles of Environmental Engineering

- Sustainable development and design principles
- Life cycle assessment and environmental impact analysis
- Integration of engineering and ecological principles

V. Water Resource Management and Conservation

- Importance of water resource management
- Techniques for water conservation and efficient use
- Innovative solutions for water treatment and purification

VI. Waste Management and Circular Economy

- Challenges of waste management and disposal
- Strategies for waste reduction, recycling, and reuse

- Introduction to the concept of circular economy

VII. Air Quality Control and Pollution Prevention

- Understanding air pollutants and their sources
- Technologies for air pollution control and monitoring
- Strategies for minimizing emissions and promoting clean air

VIII. Sustainable Urban Development

- Urbanization and its environmental impacts
- Sustainable urban planning and design principles
- Smart cities and innovative urban solutions

IX. Renewable Energy and Energy Efficiency

- Importance of transitioning to renewable energy sources
- Overview of different renewable energy technologies
- Strategies for improving energy efficiency in various sectors

X. Climate Change Adaptation and Mitigation

- Understanding the impacts of climate change
- Strategies for adapting to climate change effects
- Mitigation measures to reduce greenhouse gas emissions

XI. Policy and Governance for Environmental Engineering

- Role of government policies and regulations
- International environmental agreements and frameworks

- Community engagement and public participation

XII. Case Studies and Success Stories

- Showcase of real-world environmental engineering projects
- Examining their impacts and lessons learned
- Highlighting success stories and best practices

XIII. Future Directions and Opportunities

- Emerging trends in environmental engineering
- Challenges and opportunities in the field
- The role of individuals and organizations in shaping a sustainable future

XIV. Conclusion

- Recap of key insights and takeaways
- Call to action for fostering harmony with nature
- Encouragement for continued exploration and engagement

The urgency of environmental challenges

The urgency of environmental challenges has become increasingly apparent in recent years. The state of our planet is facing significant threats that demand immediate attention and action. Here are some key aspects that highlight the urgency of environmental challenges:

1. Climate Change: Climate change is a global crisis that is causing rising temperatures, extreme weather events, sea-level rise, and disruptions to ecosystems. The scientific consensus is clear: human activities, particularly the burning of fossil fuels, are the primary drivers of climate change. The consequences of climate change, such as more frequent and severe droughts, floods, and storms, pose significant risks to human well-being, biodiversity, and food security.

2. Biodiversity Loss: The Earth is currently experiencing an unprecedented rate of biodiversity loss. Habitat destruction, pollution, overexploitation, and climate change are pushing numerous species towards extinction. Loss of biodiversity not only disrupts ecosystems but also threatens the delicate balance that sustains life on our planet. Biodiversity loss can have cascading effects on ecosystems, leading to the collapse of vital ecological services upon which humans rely.

3. Pollution and Environmental Contamination: Pollution in various forms, including air pollution, water pollution, and soil contamination, is pervasive and detrimental to both human and environmental health. Industrial activities, improper waste management, and

the use of harmful chemicals contribute to the pollution burden. The impacts of pollution include respiratory and cardiovascular diseases, degradation of water resources, and harm to ecosystems and wildlife.

4. Depletion of Natural Resources: Unsustainable extraction and consumption of natural resources, such as freshwater, minerals, and forests, are depleting the Earth's finite resources at an alarming rate. This depletion leads to ecological imbalances, habitat destruction, and social conflicts. It threatens the availability of essential resources for future generations and compromises the well-being of both humans and ecosystems.

5. Loss of Ecosystem Services: Ecosystem services, such as clean air, clean water, fertile soil, and climate regulation, are essential for sustaining life on Earth. However, human activities have disrupted and degraded ecosystems, diminishing their ability to provide these services. The loss of ecosystem services not only impacts natural systems but also hinders our ability to thrive and adapt in a changing world.

The urgency of these environmental challenges cannot be overstated. They require immediate and collective action on a global scale. Governments, organizations, businesses, and individuals all have a role to play in addressing these challenges. It is crucial to adopt sustainable practices, transition to cleaner and renewable energy sources, promote conservation and biodiversity protection, and advocate for policy changes that prioritize environmental stewardship.

Addressing the urgency of environmental challenges is not just a responsibility; it is also an opportunity for positive change. By acting decisively and collectively, we can safeguard the planet for future generations, protect biodiversity, mitigate climate change, and create a sustainable and resilient world that thrives in harmony with nature.

The role of environmental engineering in finding sustainable solutions

Environmental engineering plays a crucial role in finding sustainable solutions to the pressing environmental challenges we face today. By applying engineering principles and scientific knowledge, environmental engineers work towards the preservation, protection, and restoration of the natural environment. Here are some key aspects that highlight the role of environmental engineering in finding sustainable solutions:

1. Environmental Impact Assessment: Environmental engineers assess the potential environmental impacts of proposed projects or activities. They evaluate the potential risks to ecosystems, air and water quality, and human health, and provide recommendations to mitigate or minimize negative impacts.

2. Pollution Control and Remediation: Environmental engineers develop and implement strategies to control and reduce pollution. They design and operate systems for wastewater treatment, air pollution control, solid waste management, and hazardous waste remediation. Through innovative technologies and practices, environmental engineers help to minimize pollution and restore contaminated environments.

3. Sustainable Infrastructure Design: Environmental engineers play a crucial role in designing infrastructure that minimizes environmental impacts. They integrate sustainable design principles into projects such as buildings, transportation systems, water supply

networks, and energy infrastructure. By considering factors such as energy efficiency, resource conservation, and ecosystem protection, environmental engineers contribute to the development of more sustainable and resilient infrastructure.

4. Resource Management and Conservation: Environmental engineers work towards the efficient and sustainable management of natural resources. They develop strategies for water resource management, including water conservation, wastewater recycling, and rainwater harvesting. They also promote sustainable practices for land use, forest management, and renewable energy production.

5. Climate Change Adaptation and Mitigation: Environmental engineers contribute to efforts to address climate change. They develop strategies to adapt to the impacts of climate change, such as designing resilient infrastructure to withstand extreme weather events. They also work on initiatives to mitigate greenhouse gas emissions through renewable energy technologies, energy efficiency measures, and carbon capture and storage.

6. Environmental Monitoring and Modeling: Environmental engineers use monitoring and modeling techniques to assess the state of the environment and predict potential impacts. They collect and analyze data on air quality, water quality, biodiversity, and other environmental indicators to inform decision-making and develop effective solutions.

7. Policy and Regulation: Environmental engineers provide technical expertise to support the development of environmental policies and regulations. They contribute to the formulation of standards and guidelines for pollution control, environmental management, and sustainable practices. They also collaborate with policymakers to ensure that

environmental considerations are integrated into decision-making processes.

Environmental engineering plays a vital role in finding sustainable solutions by combining scientific knowledge, engineering principles, and a commitment to environmental stewardship. Through their work, environmental engineers contribute to the protection of ecosystems, the preservation of natural resources, and the creation of a more sustainable and resilient future. Their efforts are essential in addressing the environmental challenges we face and ensuring the well-being of both present and future generations.

The significance of achieving harmony with nature

Achieving harmony with nature is of paramount significance for the well-being of both the planet and humanity. Here are some key reasons why it is important to strive for harmony with nature:

1. Ecosystem Balance: Nature operates in delicate balances, where all living organisms and their environments are interconnected. When these balances are disrupted, it can lead to ecological crises and cascading effects throughout the ecosystem. Achieving harmony with nature helps to maintain these balances, preserving biodiversity, and ensuring the sustainability of ecosystems.

2. Human Health and Well-being: The well-being of humans is intimately linked to the health of the natural environment. Clean air, clean water, fertile soil, and diverse ecosystems are essential for human health. By preserving and protecting nature, we safeguard the resources and ecological services necessary for our own well-being.

3. Sustainable Resource Management: Harmony with nature involves utilizing natural resources in a sustainable and responsible manner. It means recognizing the finite nature of resources and finding ways to conserve and manage them for the benefit of current and future generations. By promoting sustainable resource management, we can ensure the availability of essential resources while minimizing

environmental degradation.

4. Climate Change Mitigation: Achieving harmony with nature is closely tied to addressing climate change. By reducing greenhouse gas emissions, promoting renewable energy, and protecting forests and other carbon sinks, we can mitigate the impacts of climate change and work towards a more stable and habitable planet.

5. Conservation of Biodiversity: Biodiversity is essential for the functioning of ecosystems and provides numerous ecological, economic, and cultural benefits. Preserving biodiversity through efforts such as habitat conservation, species protection, and sustainable land management helps to maintain the intricate web of life on Earth.

6. Cultural and Spiritual Connections: Many cultures and indigenous communities have deep-rooted connections with nature, considering it sacred and recognizing the importance of living in harmony with it. Preserving and respecting these cultural and spiritual connections to nature helps to maintain diverse worldviews and promotes cultural diversity and identity.

7. Future Generations: Striving for harmony with nature is an investment in the future. By taking responsible actions today, we ensure that future generations can inherit a healthy and thriving planet. It is our duty to be good stewards of the Earth and to leave behind a sustainable legacy for the generations to come.

Ultimately, achieving harmony with nature is not only an environmental imperative but also a moral and ethical responsibility. It requires collective action, sustainable practices, and a shift in the way we interact with the natural world. By nurturing a deep respect for nature and embracing sustainable lifestyles and practices, we can create a future where humans and the natural world can coexist in balance and thrive together.

Definition and scope of environmental engineering

Environmental engineering is a branch of engineering that focuses on the application of scientific and engineering principles to address environmental challenges and promote sustainable development. It encompasses various fields and disciplines, including civil engineering, chemical engineering, mechanical engineering, and others. The scope of environmental engineering is broad and includes:

1. Environmental Assessment and Management: Environmental engineers assess the potential impacts of human activities on the environment and develop strategies to manage and mitigate those impacts. This involves conducting environmental impact assessments, developing environmental management plans, and ensuring compliance with environmental regulations and standards.

2. Water and Wastewater Management: Environmental engineers design and manage systems for the treatment and distribution of clean water and the collection and treatment of wastewater. They develop technologies and processes to ensure safe drinking water supplies, protect water resources, and manage the disposal of wastewater in an environmentally responsible manner.

3. Air Pollution Control: Environmental engineers work on the design and implementation of systems and technologies to control air pollution. They develop strategies to minimize emissions of pollutants from

industrial processes, transportation, and other sources. This includes designing and operating air pollution control systems, monitoring air quality, and developing models to predict and manage air pollution.

4. Solid and Hazardous Waste Management: Environmental engineers develop and implement strategies for the proper management and disposal of solid and hazardous waste. They design and operate systems for waste collection, recycling, and disposal. They also work on the remediation of contaminated sites and the safe handling and disposal of hazardous materials.

5. Environmental Remediation: Environmental engineers are involved in the cleanup and restoration of contaminated sites. They develop remediation plans, implement cleanup technologies, and monitor the progress of environmental remediation projects. This includes addressing soil and groundwater contamination, managing hazardous waste sites, and restoring ecosystems impacted by pollution.

6. Sustainable Infrastructure and Design: Environmental engineers play a role in the design and construction of sustainable infrastructure. They integrate environmental considerations into the planning and design of buildings, transportation systems, energy infrastructure, and other projects. This involves promoting energy efficiency, sustainable materials, and environmentally friendly construction practices.

7. Environmental Modeling and Data Analysis: Environmental engineers use modeling tools and data analysis techniques to assess environmental impacts, predict future trends, and make informed decisions. They develop computer models to simulate and predict the behavior of environmental systems, analyze large datasets to identify trends and patterns, and use statistical methods to evaluate the effectiveness of

environmental interventions.

The field of environmental engineering is dynamic and continuously evolving to address emerging environmental challenges. It requires a multidisciplinary approach, combining scientific knowledge, engineering principles, and an understanding of environmental policies and regulations. Environmental engineers play a critical role in promoting sustainable development, protecting the environment, and ensuring the well-being of current and future generations.

Interdisciplinary nature of the field

The field of environmental engineering is highly interdisciplinary, drawing upon various scientific and engineering disciplines to address complex environmental challenges. Here are some key aspects that highlight the interdisciplinary nature of the field:

1. Engineering Disciplines: Environmental engineering incorporates principles and practices from various engineering disciplines. Civil engineering plays a significant role in areas such as water and wastewater management, sustainable infrastructure design, and environmental remediation. Chemical engineering contributes expertise in areas such as air pollution control and hazardous waste management. Mechanical engineering is involved in energy efficiency, renewable energy systems, and sustainable technology development.

2. Natural Sciences: Environmental engineering relies on knowledge from the natural sciences to understand the fundamental processes and dynamics of the natural environment. This includes disciplines such as chemistry, biology, ecology, geology, and atmospheric science. Understanding the behavior of pollutants, the functioning of ecosystems, and the impacts of human activities on the natural environment requires a solid foundation in these sciences.

3. Environmental Policy and Law: Environmental engineering incorporates elements of environmental policy and law. Environmental engineers need to

have an understanding of environmental regulations, standards, and policies at the local, national, and international levels. They must consider the legal frameworks and guidelines that govern environmental protection and sustainable practices in their work.

4. Social Sciences and Economics: Environmental engineering recognizes the importance of social, economic, and human dimensions in addressing environmental challenges. Social sciences, such as sociology and psychology, provide insights into human behavior, public perceptions, and societal responses to environmental issues. Economics plays a role in evaluating the costs and benefits of environmental interventions, assessing the economic viability of sustainable projects, and incorporating economic incentives for environmental protection.

5. Data Analysis and Modeling: Environmental engineering relies on data analysis and modeling techniques to understand environmental systems, predict impacts, and inform decision-making. This involves statistical analysis, computer modeling, geographic information systems (GIS), and remote sensing. By analyzing large datasets, environmental engineers can identify patterns, trends, and correlations that help them assess environmental risks and develop effective solutions.

6. Collaboration and Interdisciplinary Research: Given the complex nature of environmental challenges, interdisciplinary collaboration is essential. Environmental engineers often collaborate with professionals from other disciplines, such as ecologists, hydrologists, urban planners, policy analysts, and social scientists. This collaboration promotes holistic approaches to problem-solving, integrating diverse perspectives and expertise to develop comprehensive and sustainable solutions.

The interdisciplinary nature of environmental engineering enables a comprehensive understanding of environmental issues and facilitates the development of innovative and integrated solutions. It recognizes that addressing environmental challenges requires a multidimensional approach that goes beyond technical engineering solutions. By integrating knowledge and expertise from various disciplines, environmental engineers are better equipped to tackle complex environmental problems and promote sustainable practices.

The importance of systems thinking and holistic approaches

Systems thinking and holistic approaches are of utmost importance in the field of environmental engineering. They enable a comprehensive understanding of environmental challenges and help develop effective and sustainable solutions. Here's why systems thinking and holistic approaches are crucial:

1. Understanding Interconnections: Environmental issues are often interconnected and operate within complex systems. Systems thinking allows environmental engineers to analyze the relationships and interdependencies between different components of a system. By understanding these interconnections, they can identify the root causes of environmental problems and consider the potential consequences of interventions on the entire system.

2. Addressing Complexity: Environmental challenges are multifaceted and involve a range of social, economic, and ecological factors. Holistic approaches take into account the various dimensions of a problem and consider the broader context in which it exists. This helps environmental engineers identify and address the underlying complexities that contribute to the problem, leading to more effective and sustainable solutions.

3. Preventing Unintended Consequences: Focusing solely on isolated aspects of an environmental problem can lead to unintended consequences. For example, a solution that reduces air pollution in one area may

inadvertently increase water pollution or contribute to other environmental issues. Holistic approaches help anticipate and mitigate potential negative impacts by considering the broader implications of interventions on the entire system.

4. Promoting Sustainability: Sustainability is a central goal in environmental engineering. Systems thinking and holistic approaches emphasize the long-term viability of solutions by considering social, economic, and environmental factors simultaneously. This helps ensure that interventions address not only immediate needs but also contribute to the overall well-being of ecosystems, communities, and future generations.

5. Integrated Decision-Making: Environmental engineers often work in collaboration with policymakers, stakeholders, and other professionals. Systems thinking facilitates integrated decision-making processes by incorporating diverse perspectives, expertise, and values. It helps identify trade-offs, synergies, and opportunities for win-win solutions that benefit multiple stakeholders and promote sustainable outcomes.

6. Adaptive Management: Environmental challenges are dynamic, and the effectiveness of solutions may evolve over time. Holistic approaches promote adaptive management by considering the feedback loops and iterative processes within a system. This allows environmental engineers to continuously monitor, evaluate, and adjust their interventions based on changing circumstances and new information.

7. Transdisciplinary Collaboration: Systems thinking and holistic approaches encourage collaboration across disciplines and sectors. Environmental challenges often require expertise from multiple fields, such as science, engineering, social sciences, and policy. By engaging in transdisciplinary collaboration, environmental

engineers can leverage diverse knowledge and perspectives to develop comprehensive and innovative solutions.

In summary, systems thinking and holistic approaches are essential in environmental engineering to address the complexities of environmental challenges and develop sustainable solutions. By considering the interconnections, complexities, and broader contexts of environmental issues, environmental engineers can effectively tackle problems, promote sustainability, and work towards a harmonious relationship between humans and the natural world.

Overview of key environmental systems (e.g., water, air, land)

Environmental systems encompass various components of the Earth's natural environment, including water, air, and land. Here's an overview of key environmental systems:

1. Water Systems:
   - Surface Water: Includes rivers, lakes, reservoirs, and oceans. It serves as a crucial resource for drinking water, agriculture, industry, and ecosystems.
   - Groundwater: Refers to water stored underground in aquifers. It plays a vital role in supplying drinking water and supporting ecosystems.
   - Water Cycle: The continuous movement of water between the Earth's surface and the atmosphere through processes such as evaporation, condensation, and precipitation.

2. Air Systems:
   - Atmosphere: The layer of gases surrounding the Earth, including nitrogen, oxygen, carbon dioxide, and other trace gases. It supports life, regulates temperature, and affects weather patterns.
   - Air Quality: The measure of pollutants, such as particulate matter, nitrogen oxides, sulfur dioxide, and volatile organic compounds, present in the air. Maintaining clean air is

crucial for human health and ecosystem well-being.

3. Land Systems:
    - Soil: A complex mixture of minerals, organic matter, water, air, and organisms. It provides the foundation for plant growth, supports ecosystems, and plays a critical role in nutrient cycling.
    - Ecosystems: Diverse habitats and communities of organisms, including forests, grasslands, wetlands, and deserts. They provide essential services such as carbon sequestration, water filtration, and habitat for biodiversity.
    - Land Use: The human utilization of land for various purposes, including agriculture, urban development, mining, and infrastructure. Sustainable land use practices aim to balance human needs with environmental protection.

4. Coastal and Marine Systems:
    - Coastal Zones: Areas where land and sea meet, including beaches, estuaries, and mangroves. They are ecologically rich and provide important habitat for diverse species.
    - Oceans: Vast bodies of saltwater covering a significant portion of the Earth's surface. Oceans regulate climate, support marine life, and provide resources such as food and energy.

5. Biodiversity:
    - Biological Diversity: The variety of life forms, including plants, animals, and microorganisms, in different ecosystems. Biodiversity is crucial for ecosystem stability, food security, and the provision of ecosystem services.

Understanding and managing these key environmental systems

is essential for sustainable development and the preservation of Earth's natural resources. Environmental engineers and scientists work to monitor and protect these systems, develop sustainable management strategies, and promote practices that maintain the health and balance of these interconnected components. By safeguarding water, air, land, and biodiversity, we can ensure a sustainable and thriving planet for future generations.

Examination of global environmental challenges (e.g., climate change, pollution, resource depletion)

Global environmental challenges pose significant threats to the planet and require urgent attention. Here is an examination of some key global environmental challenges:

1. Climate Change: Climate change is a defining challenge of our time. It is primarily driven by human activities, particularly the burning of fossil fuels and deforestation, resulting in increased greenhouse gas emissions. Climate change leads to rising global temperatures, sea-level rise, altered precipitation patterns, extreme weather events, and disruptions to ecosystems. Its impacts are far-reaching, affecting agriculture, water resources, public health, biodiversity, and human livelihoods.

2. Air Pollution: Air pollution, mainly caused by industrial activities, transportation, and the burning of fossil fuels, poses significant health and environmental risks. It leads to the release of pollutants such as particulate matter, nitrogen oxides, sulfur dioxide, and volatile organic compounds. Air pollution contributes to respiratory and cardiovascular diseases, damages ecosystems, and deteriorates air quality in urban areas and regions downwind of pollution sources.

3. Water Scarcity and Water Pollution: Access to clean and safe water is a fundamental human right, yet

water scarcity and water pollution are global challenges. Growing population, industrialization, and agricultural practices contribute to water scarcity in many regions. Additionally, contamination from pollutants, chemicals, and inadequate sanitation systems leads to water pollution, endangering ecosystems and human health.

4. Biodiversity Loss and Habitat Destruction: The loss of biodiversity is accelerating at an alarming rate due to factors such as habitat destruction, deforestation, pollution, climate change, and overexploitation. Biodiversity loss disrupts ecosystems, reduces resilience, and threatens the delicate balance that supports life on Earth. It has profound implications for food security, ecosystem services, and the preservation of unique species.

5. Land Degradation and Deforestation: Unsustainable land use practices, including deforestation, overgrazing, and unsustainable agriculture, contribute to land degradation. These activities lead to soil erosion, desertification, loss of fertile land, and habitat destruction. Land degradation threatens food production, exacerbates climate change, and compromises the sustainability of ecosystems.

6. Resource Depletion: Unsustainable consumption patterns and the depletion of natural resources, such as freshwater, minerals, and fossil fuels, pose significant challenges. Overexploitation of resources, coupled with inefficient use and waste generation, results in environmental degradation, ecological imbalances, and conflicts over scarce resources. Resource depletion also raises concerns about future generations' ability to meet their needs.

7. Waste Management: Improper waste management and the generation of vast amounts of solid and hazardous waste contribute to pollution, soil and

water contamination, and negative health impacts. The improper disposal of waste and the lack of recycling and waste reduction practices further strain natural resources and exacerbate environmental challenges.

Addressing these global environmental challenges requires collective action, transformative changes in policy and behavior, and the adoption of sustainable practices. It necessitates efforts to reduce greenhouse gas emissions, transition to renewable energy sources, promote circular economy principles, protect ecosystems, conserve biodiversity, enhance waste management systems, and foster sustainable land and water management practices. By addressing these challenges, we can work towards a more sustainable and resilient future for our planet and future generations.

Impacts of human activities on the environment

Human activities have had significant impacts on the environment, altering ecosystems, depleting natural resources, and contributing to various environmental challenges. Here are some key impacts of human activities on the environment:

1. Climate Change: The burning of fossil fuels, deforestation, and industrial activities have led to a substantial increase in greenhouse gas emissions, resulting in climate change. Rising global temperatures, altered precipitation patterns, sea-level rise, and extreme weather events are some of the consequences of these activities. Climate change disrupts ecosystems, affects biodiversity, threatens food security, and poses risks to human health and livelihoods.

2. Deforestation: Large-scale deforestation, primarily driven by agriculture, logging, and urbanization, has had profound impacts on ecosystems. Deforestation leads to habitat loss, biodiversity decline, soil erosion, and changes in local and regional climates. It also releases significant amounts of carbon dioxide into the atmosphere, contributing to climate change.

3. Loss of Biodiversity: Human activities, including habitat destruction, overexploitation, pollution, and the introduction of invasive species, have resulted in a rapid loss of biodiversity. Species extinction rates are currently much higher than the natural background rate. The loss of biodiversity disrupts ecosystems, reduces resilience, and threatens the stability of ecological processes.

4. Pollution: Various forms of pollution, such as air pollution, water pollution, and soil contamination, are the consequences of human activities. Industrial emissions, vehicle emissions, improper waste disposal, and the use of harmful chemicals release pollutants into the environment, harming human health and ecosystems. Pollution impacts air quality, water resources, aquatic life, soil fertility, and biodiversity.

5. Resource Depletion: Human activities have led to the overexploitation and depletion of natural resources, including freshwater, minerals, and fossil fuels. Unsustainable extraction and consumption patterns, coupled with population growth and industrialization, have put immense strain on natural resources. This depletion threatens the availability of essential resources, contributes to ecological imbalances, and drives environmental degradation.

6. Land Degradation: Unsustainable land use practices, such as deforestation, overgrazing, and inappropriate agriculture, have caused soil erosion, desertification, and land degradation. These activities degrade fertile soil, compromise agricultural productivity, and lead to habitat loss. Land degradation also contributes to climate change through the release of stored carbon in soils.

7. Water Scarcity: Human activities, including excessive water consumption, pollution, and poor water management practices, have led to water scarcity in many regions. Overexploitation of water resources, contamination of water bodies, and the alteration of natural water cycles have resulted in water scarcity, impacting ecosystems, agriculture, and human communities.

These impacts highlight the need for sustainable practices, conservation efforts, and environmental stewardship. It is crucial to adopt sustainable development strategies, promote resource efficiency, reduce greenhouse gas emissions, protect ecosystems, conserve biodiversity, and implement responsible waste and water management practices. By addressing the impacts of human activities on the environment, we can strive towards a more sustainable and resilient future.

Sustainable development and design principles

Sustainable development and design principles aim to balance social, economic, and environmental considerations to meet the needs of the present without compromising the ability of future generations to meet their own needs. These principles guide the development of practices, policies, and designs that promote long-term sustainability and minimize negative impacts on the environment. Here are some key sustainable development and design principles:

1. Environmental Protection: Sustainable development emphasizes the protection and preservation of the natural environment. This involves minimizing pollution and waste generation, conserving resources, preserving biodiversity, and mitigating climate change impacts. Design principles focus on reducing the environmental footprint of products, buildings, and infrastructure through efficient resource use, responsible waste management, and the use of renewable energy sources.

2. Social Equity: Sustainable development recognizes the importance of social equity and ensuring that development benefits all members of society. Design principles strive for inclusivity, accessibility, and equal access to resources, services, and opportunities. This includes considerations of social justice, fair distribution of benefits and burdens, and the empowerment of marginalized communities.

3. Economic Viability: Sustainable development requires economic systems that are financially viable and

support long-term prosperity. Design principles promote economic efficiency, innovation, and the integration of sustainability into business practices. This includes adopting circular economy principles, promoting sustainable consumption and production patterns, and considering the life cycle costs and benefits of products and projects.

4. Integration and Collaboration: Sustainable development principles encourage integration and collaboration across disciplines, sectors, and stakeholders. Design principles promote multidisciplinary approaches, involving input from diverse perspectives, expertise, and knowledge. This collaborative approach ensures that decisions are informed, holistic, and consider the interconnectedness of social, economic, and environmental systems.

5. Resilience and Adaptability: Sustainable development and design principles emphasize building resilience and adaptability to environmental and social changes. This involves considering future scenarios, anticipating risks and uncertainties, and designing for flexibility and robustness. Resilient designs and strategies enable communities and systems to withstand and recover from shocks and stresses, such as natural disasters and climate change impacts.

6. Life Cycle Thinking: Sustainable development and design principles incorporate life cycle thinking, which considers the environmental, economic, and social impacts of a product, service, or project throughout its entire life cycle. This involves assessing the environmental and social implications of raw material extraction, production, use, and disposal. Life cycle thinking helps identify opportunities for improvement, optimize resource use, and minimize negative impacts.

7. Education and Awareness: Sustainable development principles emphasize the importance of education,

awareness, and capacity building. Design principles prioritize education and public engagement to promote sustainable behavior, encourage responsible decision-making, and foster a culture of sustainability. This includes raising awareness about environmental issues, promoting sustainable lifestyles, and enhancing knowledge and skills related to sustainable development.

By integrating these sustainable development and design principles, we can strive for a more sustainable and resilient future. They guide the development of innovative solutions, inform policy decisions, and shape the way we design and interact with the built and natural environment. Embracing these principles is essential for achieving a balance between environmental protection, social equity, and economic well-being.

Life cycle assessment and environmental impact analysis

Life cycle assessment (LCA) and environmental impact analysis are methodologies used to evaluate the environmental impacts of a product, process, or service throughout its entire life cycle. Here's an overview of these approaches:

Life Cycle Assessment (LCA): Life cycle assessment is a systematic and comprehensive method for assessing the environmental impacts of a product, process, or service from "cradle to grave" or "cradle to cradle." It considers all stages of the life cycle, including raw material extraction, manufacturing, transportation, use, and end-of-life disposal or recycling. LCA quantifies and evaluates various environmental indicators, such as energy use, greenhouse gas emissions, water consumption, waste generation, and resource depletion.

The LCA methodology typically involves the following steps:

1. Goal and Scope Definition: Clearly defining the purpose and boundaries of the LCA study, including the system under analysis, functional unit, and environmental impact categories to be assessed.
2. Inventory Analysis: Collecting data on the inputs, outputs, and emissions associated with each life cycle stage. This involves gathering information on energy consumption, material use, emissions, waste generation, and transportation.
3. Impact Assessment: Evaluating the potential environmental impacts based on the collected inventory

data. This step involves assessing the impacts in different categories such as climate change, human health, ecosystem quality, and resource depletion.

4. Interpretation: Interpreting the LCA results, drawing conclusions, and communicating the findings to stakeholders. This step involves considering the limitations, uncertainties, and assumptions of the study and identifying improvement opportunities based on the results.

Environmental Impact Analysis: Environmental impact analysis is a broader term that encompasses various methods used to assess the environmental impacts of a specific project, policy, or activity. It may include methods such as environmental impact assessment (EIA), which is a systematic process for evaluating the potential environmental effects of proposed projects or activities before they are implemented. Environmental impact analysis also considers the social and economic dimensions of the impacts, alongside the environmental aspects.

Environmental impact analysis typically involves the following steps:

1. Scoping: Defining the boundaries and objectives of the analysis, identifying the potential environmental impacts to be assessed, and establishing the baseline conditions for comparison.

2. Impact Identification: Identifying and characterizing the potential environmental impacts associated with the project or activity. This may involve assessing impacts on air quality, water resources, biodiversity, land use, noise pollution, and social aspects.

3. Impact Prediction and Evaluation: Assessing the magnitude, extent, and significance of the identified impacts. This step involves analyzing the potential consequences of the project or activity on the environment and comparing them to established

criteria, standards, or regulations.

4. Mitigation and Monitoring: Proposing measures to mitigate or minimize the identified impacts and monitoring their implementation and effectiveness throughout the project's life cycle.

Both LCA and environmental impact analysis provide valuable tools for decision-making, policy development, and identifying opportunities for environmental improvement. They help identify hotspots, prioritize actions, and inform the design of more sustainable products, processes, and policies. These methodologies contribute to the understanding of the environmental implications of human activities and support efforts to minimize negative environmental impacts.

Integration of engineering and ecological principles

The integration of engineering and ecological principles is crucial for promoting sustainable development and addressing environmental challenges. By combining engineering and ecological perspectives, we can design solutions that not only meet human needs but also maintain and enhance the health of ecosystems. Here are some key aspects of the integration of engineering and ecological principles:

1. Ecosystem-based Approaches: Engineering projects can benefit from adopting ecosystem-based approaches that consider the ecological functions and services provided by natural systems. By understanding and incorporating the natural processes and functions of ecosystems into engineering designs, we can develop more sustainable and resilient solutions. For example, using natural wetlands for water filtration or designing green infrastructure that mimics natural processes can help mitigate the impacts of urbanization on water quality and quantity.

2. Green Infrastructure and Nature-Based Solutions: Integrating engineering and ecological principles involves incorporating green infrastructure and nature-based solutions into designs. This includes using vegetation, natural drainage systems, and permeable surfaces to manage stormwater, mitigate heat island effects, enhance biodiversity, and improve the overall quality of the built environment. By integrating natural

elements, we can create sustainable and aesthetically pleasing infrastructure that benefits both humans and the environment.

3. Sustainable Land Use Planning: Engineering and ecological principles can be integrated into land use planning processes to minimize negative environmental impacts. This involves considering ecological factors such as habitat connectivity, biodiversity conservation, and ecosystem services when designing urban or rural landscapes. By prioritizing green spaces, preserving natural areas, and promoting compact and mixed-use development, we can create more sustainable and ecologically sensitive communities.

4. Environmental Restoration and Remediation: Engineering techniques and ecological principles can be combined in environmental restoration and remediation projects. This includes restoring degraded ecosystems, rehabilitating contaminated sites, and implementing habitat restoration initiatives. Engineering solutions can be used to address specific environmental challenges, while ecological principles guide the selection and implementation of restoration strategies that promote the recovery of ecosystems.

5. Environmental Monitoring and Assessment: Integrating engineering and ecological principles involves the use of monitoring and assessment techniques to understand the impacts of engineering projects on the environment. By incorporating ecological indicators and measurements alongside engineering metrics, we can better evaluate the effectiveness and sustainability of projects. This enables adaptive management, where feedback from ecological monitoring guides adjustments and improvements to engineering practices.

6. Collaboration and Interdisciplinary Research: Integrating engineering and ecological principles

requires collaboration among professionals from different disciplines, including engineers, ecologists, planners, and policymakers. Interdisciplinary research and collaboration foster innovation, promote holistic problem-solving, and contribute to the development of sustainable and integrated solutions. By leveraging diverse expertise, we can bridge the gap between engineering and ecology and create more effective and environmentally conscious designs and practices.

The integration of engineering and ecological principles provides a pathway towards more sustainable and environmentally friendly approaches to development. By considering ecological processes, functions, and services alongside engineering considerations, we can design and implement solutions that support both human well-being and the health of ecosystems. This approach is vital for achieving a harmonious relationship between human activities and the natural environment.

Importance of water resource management

Water resource management is of paramount importance for the sustainable development and well-being of societies and ecosystems. Here are some key reasons highlighting the significance of water resource management:

1. Access to Clean Water: Water is essential for human survival, health, and sanitation. Effective water resource management ensures reliable access to clean and safe water for drinking, hygiene, and sanitation purposes. It involves managing water sources, treatment systems, distribution networks, and wastewater treatment facilities to ensure the provision of safe and accessible water to communities.

2. Sustainable Agriculture: Water is a critical input in agriculture, supporting crop irrigation and livestock production. Proper water resource management helps optimize water use in agriculture, promoting efficient irrigation techniques, soil moisture management, and water-saving practices. It helps ensure food security, improve agricultural productivity, and reduce water wastage.

3. Ecosystem Health: Healthy ecosystems rely on well-managed water resources. Proper water management is crucial for maintaining aquatic ecosystems, wetlands, and biodiversity. It involves maintaining sufficient water flows, preserving natural habitats, and mitigating the impacts of human activities on aquatic ecosystems. Sustainable water resource management helps protect and restore ecosystems, supporting the survival of

various plant and animal species.

4. Climate Change Adaptation: Climate change affects water availability, distribution, and quality. Effective water resource management is essential for adapting to the impacts of climate change. It involves assessing water vulnerabilities, implementing water-saving measures, promoting water reuse and recycling, and developing climate-resilient infrastructure. By managing water resources in a sustainable and adaptive manner, communities can better cope with the challenges posed by climate change.

5. Disaster Risk Reduction: Proper water resource management plays a crucial role in reducing the risks associated with water-related disasters such as floods and droughts. It involves implementing flood management strategies, developing early warning systems, and promoting drought preparedness measures. Well-managed water resources can help minimize the impacts of extreme weather events, protect communities, and reduce economic losses.

6. Economic Development: Water is a valuable economic resource, supporting various industries and economic activities. Effective water resource management contributes to economic development by ensuring sufficient water supplies for industrial processes, energy production, tourism, and other sectors. It also involves balancing competing water demands, promoting water-efficient technologies, and considering the economic value of water in decision-making processes.

7. International Cooperation: Water resources often transcend political boundaries, leading to shared water management challenges. Effective water resource management requires international cooperation and transboundary collaboration to ensure equitable and sustainable use of shared water resources. Collaborative efforts can help prevent conflicts, foster dialogue, and

promote the equitable sharing of water benefits among different countries and communities.

In summary, water resource management is essential for ensuring access to clean water, supporting sustainable agriculture, protecting ecosystems, adapting to climate change, reducing disaster risks, promoting economic development, and fostering international cooperation. By managing water resources in a sustainable and integrated manner, we can address the water-related challenges of today and safeguard this precious resource for future generations.

Techniques for water conservation and efficient use

Water conservation and efficient use are crucial for sustainable water resource management. Here are some techniques and practices that can help conserve water and improve water use efficiency:

1. Efficient Irrigation: Implementing efficient irrigation practices can significantly reduce water usage in agriculture. Techniques such as drip irrigation, precision sprinklers, and soil moisture sensors help deliver water directly to plant roots, minimizing evaporation and runoff. Timely irrigation scheduling and adjusting irrigation rates based on weather conditions and crop needs can further optimize water use.

2. Water-Efficient Landscaping: Designing and maintaining water-efficient landscapes can save significant amounts of water. This includes planting native and drought-tolerant vegetation, using mulch to retain moisture in the soil, and grouping plants with similar water needs together. Employing smart irrigation controllers that adjust watering based on weather conditions and soil moisture levels can also contribute to water savings.

3. Water-Efficient Appliances and Fixtures: Installing water-efficient appliances and fixtures in homes and businesses can reduce water consumption. Low-flow toilets, showerheads, and faucets can significantly

lower water usage without compromising functionality. Energy-efficient washing machines and dishwashers also help conserve water by optimizing water usage during cycles.

4. Rainwater Harvesting: Capturing and storing rainwater for later use is an effective way to conserve water. Rainwater can be collected from rooftops and directed into storage tanks or underground cisterns. This harvested water can be used for landscape irrigation, toilet flushing, and other non-potable purposes, reducing reliance on freshwater sources.

5. Greywater Recycling: Recycling and reusing greywater, which is water from sources like sinks, showers, and laundry, can contribute to water conservation. Greywater systems treat and filter this water for non-potable uses such as irrigation or flushing toilets. Proper treatment and safe handling ensure that recycled greywater does not pose health risks.

6. Leak Detection and Repair: Regularly inspecting and repairing leaks in plumbing systems is essential to prevent water waste. Leaky faucets, pipes, and toilets can result in significant water loss over time. Monitoring water bills and conducting periodic inspections can help identify and address leaks promptly.

7. Public Awareness and Education: Raising public awareness about the importance of water conservation and providing education on water-saving practices can have a significant impact. Promoting water conservation through public campaigns, educational programs, and incentives encourages individuals and communities to adopt water-saving behaviors and make conscious choices about water use.

8. Industrial and Commercial Water Management: Industries and commercial facilities can implement water management strategies to reduce water

consumption. These may include optimizing production processes to minimize water usage, recycling and reusing water within operations, and implementing water-efficient technologies and practices.

By implementing these water conservation and efficient use techniques, individuals, communities, businesses, and industries can contribute to water sustainability and help alleviate water stress. Conserving water not only ensures its availability for future generations but also helps protect ecosystems, maintain water quality, and address the challenges posed by population growth and climate change.

Innovative solutions for water treatment and purification

Innovative solutions for water treatment and purification play a critical role in ensuring access to safe and clean water. Here are some examples of innovative technologies and approaches in water treatment:

1. Advanced Filtration Systems: Advanced filtration systems, such as membrane filtration and nanofiltration, use specialized membranes with tiny pores to remove contaminants from water. These technologies can effectively remove suspended solids, bacteria, viruses, and other pollutants, providing high-quality treated water.

2. Ultraviolet (UV) Disinfection: UV disinfection involves using ultraviolet light to kill or inactivate microorganisms present in water. It is a chemical-free and environmentally friendly method that is highly effective against bacteria, viruses, and protozoa. UV disinfection is commonly used as a final step in water treatment to ensure the elimination of any remaining pathogens.

3. Reverse Osmosis (RO): Reverse osmosis is a membrane-based water purification process that uses pressure to force water molecules through a semipermeable membrane, leaving behind contaminants. It is highly effective in removing salts, minerals, heavy metals, and other dissolved substances from water, making it suitable for drinking water production and

desalination.

4. Electrocoagulation: Electrocoagulation is an electrochemical process that uses electrical current to destabilize and coagulate suspended particles, colloids, and dissolved pollutants in water. The coagulated particles form larger flocs, which can be easily separated from the water. This method is effective in removing heavy metals, oils, dyes, and other contaminants.

5. Advanced Oxidation Processes (AOPs): AOPs involve the use of powerful oxidants, such as ozone, hydrogen peroxide, or ultraviolet light, to break down and degrade organic and inorganic pollutants in water. These processes generate highly reactive hydroxyl radicals that can destroy a wide range of contaminants, including pharmaceuticals, pesticides, and emerging pollutants.

6. Hybrid Treatment Systems: Hybrid treatment systems combine multiple treatment processes to achieve comprehensive water purification. These systems often incorporate a combination of physical, chemical, and biological treatment methods to address specific water quality challenges effectively. Examples include membrane bioreactors (MBRs), which combine membrane filtration with biological treatment, and integrated advanced oxidation and filtration systems.

7. Water Reuse and Desalination: Water reuse and desalination technologies have become increasingly important in water-scarce regions. Desalination processes, such as reverse osmosis and multi-stage flash distillation, remove salt and other impurities from seawater or brackish water, making it suitable for drinking or irrigation. Water reuse technologies treat wastewater to produce reclaimed water for non-potable uses like irrigation, industrial processes, or groundwater recharge.

8. Smart Water Management Systems: Smart water

management systems leverage digital technologies, sensors, and data analytics to optimize water treatment and distribution processes. These systems enable real-time monitoring of water quality, consumption patterns, and system performance. By analyzing data and using predictive algorithms, smart water management systems can detect leaks, optimize water usage, and improve overall efficiency.

Innovative solutions for water treatment and purification continue to evolve, driven by advances in science, engineering, and technology. These solutions are essential for addressing water scarcity, improving water quality, and ensuring access to safe and clean water for communities around the world.

Challenges of waste management and disposal

Waste management and disposal present significant challenges that impact both the environment and human health. Here are some key challenges associated with waste management:

1. Increasing Waste Generation: Rapid population growth, urbanization, and changing consumption patterns have led to a substantial increase in waste generation worldwide. Managing the growing volume of waste poses challenges in terms of collection, transportation, and disposal.

2. Inadequate Infrastructure: Many regions, particularly in developing countries, lack the necessary infrastructure for efficient waste management. Insufficient waste collection systems, inadequate landfill facilities, and limited recycling and treatment facilities contribute to improper waste disposal practices.

3. Environmental Pollution: Improper waste disposal can lead to environmental pollution. Dumping waste in open areas or bodies of water can contaminate soil, water, and air. Hazardous waste, such as chemicals, electronic waste, and medical waste, pose particular risks to ecosystems and human health if not managed and disposed of properly.

4. Landfill Overflow and Space Constraints: Landfills are the most common method of waste disposal in many parts of the world. However, landfills have limited capacity, and as waste generation increases, landfills can quickly reach their maximum capacity. Finding suitable locations for new landfills can be challenging due to

environmental concerns, land scarcity, and community opposition.

5. Health Risks and Public Health Concerns: Improper waste management can have detrimental effects on public health. Open dumping and inadequate waste treatment can result in the spread of diseases, contamination of water sources, and the proliferation of disease-carrying vectors like insects and rodents. Waste workers, such as scavengers and waste collectors, are also exposed to health risks due to the lack of proper safety measures.

6. Lack of Recycling and Resource Recovery: Inefficient waste management practices often result in the underutilization of resources and missed opportunities for recycling and recovery. Many valuable materials that could be recycled or reused end up in landfills, wasting resources and contributing to environmental degradation.

7. Financial and Economic Constraints: Implementing effective waste management systems requires substantial financial resources. Funding limitations and competing priorities often hinder the development of comprehensive waste management infrastructure, especially in economically disadvantaged regions. Lack of financial resources can impede waste collection, recycling programs, and the implementation of advanced waste treatment technologies.

8. Awareness and Behavior Change: Encouraging proper waste disposal practices and promoting responsible consumer behavior pose ongoing challenges. Education and awareness campaigns are needed to foster behavior change, encourage waste reduction, and promote recycling and proper waste segregation.

Addressing these challenges requires a multi-faceted approach, including improved waste management infrastructure, implementation of appropriate technologies, policy reforms, public participation, and behavioral change. It is crucial to prioritize waste reduction, recycling, and resource recovery to minimize environmental impacts and move towards a more sustainable and circular economy. Additionally, collaboration between governments, communities, industries, and waste management entities is essential to finding innovative and sustainable solutions to the challenges of waste management and disposal.

Strategies for waste reduction, recycling, and reuse

Strategies for waste reduction, recycling, and reuse play a crucial role in promoting sustainable waste management and minimizing the environmental impact of waste. Here are some key strategies that can be implemented:

1. Source Reduction: Source reduction focuses on minimizing waste generation at its source by adopting practices that reduce the quantity or toxicity of materials used. This can include implementing efficient manufacturing processes, reducing packaging materials, promoting product design for durability and repairability, and encouraging the use of reusable products.

2. Recycling: Recycling involves the collection, sorting, and processing of materials to create new products. It helps conserve natural resources, reduce energy consumption, and minimize waste sent to landfills. Implementing effective recycling programs involves providing accessible recycling bins, educating the public about recycling practices, and establishing partnerships with recycling facilities.

3. Composting: Composting is the process of decomposing organic waste, such as food scraps and yard trimmings, into nutrient-rich compost. Composting reduces the amount of organic waste sent to landfills and produces a valuable soil amendment. Encouraging composting at home, promoting community composting programs,

and incorporating composting in municipal waste management systems are effective strategies.

4. Waste-to-Energy Conversion: Waste-to-energy technologies convert non-recyclable waste into energy, such as heat or electricity. Processes like incineration, gasification, and anaerobic digestion can help recover energy from waste while minimizing its volume and environmental impact. These technologies should be implemented with strict emission controls to mitigate environmental and health risks.

5. Product Reuse and Repair: Promoting product reuse and repair extends the lifespan of products, reducing the need for new production and minimizing waste. Encouraging consumers to buy used or refurbished goods, supporting repair and refurbishment services, and creating platforms for sharing and swapping items are effective strategies for promoting product reuse.

6. Extended Producer Responsibility (EPR): EPR is a policy approach that holds producers responsible for the end-of-life management of their products. It encourages product design for recyclability, supports recycling infrastructure development, and promotes the collection and recycling of products at the end of their life. EPR programs can be implemented for various products, including electronics, packaging, and hazardous materials.

7. Waste Reduction in Businesses and Industries: Businesses and industries can implement waste reduction strategies by adopting cleaner production techniques, optimizing material and energy use, and implementing waste management programs. This includes reducing packaging waste, implementing efficient production processes, and exploring opportunities for by-product reuse or recycling.

8. Education and Public Awareness: Raising awareness and educating the public about the importance of waste

reduction, recycling, and reuse is crucial for promoting behavioral change. Educational campaigns, community outreach programs, and school initiatives can help inform individuals about proper waste management practices and encourage sustainable consumption habits.

By implementing these strategies, we can reduce waste generation, conserve resources, minimize environmental impacts, and move towards a more circular economy. It requires collaboration among individuals, businesses, governments, and waste management entities to create an environment that supports waste reduction, recycling, and reuse practices.

Introduction to the concept of circular economy

The concept of a circular economy represents a shift from the traditional linear economic model of "take-make-dispose" to a more sustainable and regenerative approach. In a circular economy, resources are used efficiently, waste is minimized, and materials are continuously circulated and reused, creating a closed-loop system. The aim is to decouple economic growth from resource consumption and environmental degradation.

In a circular economy, products and materials are designed to be durable, repairable, and recyclable. Instead of being discarded at the end of their life, products are recovered, repaired, or remanufactured to extend their lifespan. Materials and components are recycled or reintegrated into new products or production processes, reducing the need for virgin resources.

The circular economy concept is based on three key principles:

1. Design for Circularity: Products and systems are designed with the intent of maximizing their lifecycle and minimizing waste. This includes considering factors such as material selection, durability, ease of repair, and recyclability. Designing for circularity ensures that products can be easily disassembled, components can be recovered, and materials can be recycled or repurposed.

2. Resource Efficiency and Closed-Loop Systems: The circular economy aims to optimize resource use and minimize waste generation. This involves maximizing

the efficiency of resource extraction, manufacturing processes, and product use. It also promotes closed-loop systems where materials are continuously circulated and reused, reducing the need for virgin resources and minimizing environmental impacts.

3. Collaboration and Systems Thinking: Achieving a circular economy requires collaboration among stakeholders across the value chain. This includes cooperation between producers, consumers, businesses, policymakers, and waste management entities. Systems thinking is crucial to understanding the interconnectedness of various elements within the economy and identifying opportunities for collaboration and innovation.

The benefits of transitioning to a circular economy are numerous. It reduces resource depletion, decreases waste generation, and lowers greenhouse gas emissions. It also stimulates innovation, creates new business opportunities, and promotes job creation. By decoupling economic growth from resource consumption, the circular economy offers a sustainable and resilient framework for addressing environmental challenges and fostering a more prosperous and equitable society.

Transitioning to a circular economy requires a holistic and systemic approach, involving changes in production methods, consumption patterns, waste management systems, and policy frameworks. It calls for a collective effort to redesign systems, rethink business models, and embrace sustainable practices that enable the efficient use of resources and the preservation of the natural environment.

Understanding air pollutants and their sources

Air pollutants are substances in the air that can have harmful effects on human health, the environment, and the overall quality of the air we breathe. These pollutants can come from both natural and human-made sources. Understanding the different types of air pollutants and their sources is essential for effective air quality management. Here are some common air pollutants and their sources:

1. Particulate Matter (PM): Particulate matter refers to tiny solid or liquid particles suspended in the air. PM can be categorized based on their size: PM10 (particles with a diameter of 10 micrometers or less) and PM2.5 (particles with a diameter of 2.5 micrometers or less). Sources of PM include combustion processes (such as fossil fuel combustion in vehicles, power plants, and industrial processes), industrial emissions, construction activities, agricultural operations, and natural sources like dust and wildfires.

2. Nitrogen Oxides (NOx): Nitrogen oxides are a group of gases that are formed during the combustion of fossil fuels, primarily in vehicles, power plants, and industrial processes. The main components of NOx are nitric oxide (NO) and nitrogen dioxide (NO2). NOx emissions contribute to the formation of ground-level ozone and particulate matter and can also have direct health effects.

3. Sulfur Dioxide (SO2): Sulfur dioxide is a gas produced by the combustion of fossil fuels, particularly those containing sulfur, such as coal and oil. Major sources

of SO_2 emissions include power plants, industrial processes (such as smelting and refining), and residential heating. SO_2 can contribute to the formation of acid rain and has respiratory and cardiovascular health impacts.

4. Carbon Monoxide (CO): Carbon monoxide is a colorless and odorless gas produced by incomplete combustion of fossil fuels. Major sources of CO include vehicle emissions, industrial processes, and residential heating. Exposure to high levels of CO can lead to adverse health effects, particularly on the cardiovascular system.

5. Volatile Organic Compounds (VOCs): VOCs are organic chemicals that can easily vaporize at room temperature. They are emitted from various sources, including vehicle emissions, industrial processes, solvents, and chemical products. VOCs play a significant role in the formation of ground-level ozone and contribute to air pollution and indoor air quality issues.

6. Ozone (O_3): Ground-level ozone is not emitted directly but is formed by chemical reactions between nitrogen oxides and volatile organic compounds in the presence of sunlight. It is a major component of smog. Ozone is harmful to human health, especially to the respiratory system, and can also have adverse effects on vegetation and ecosystems.

7. Hazardous Air Pollutants (HAPs): HAPs, also known as air toxics, are a group of pollutants that are known or suspected to cause serious health effects, including cancer and other chronic health conditions. They can be emitted from industrial processes, such as chemical manufacturing, power plants, and waste incineration.

It is important to note that the sources and amounts of air pollutants can vary depending on geographical location, industrial activities, transportation patterns, and local climate conditions. Understanding the sources of air pollutants helps in developing effective emission control strategies, implementing regulations and policies, and promoting cleaner technologies to reduce the impact of air pollution on human health and the environment.

Technologies for air pollution control and monitoring

Technologies for air pollution control and monitoring play a crucial role in managing and reducing the impact of air pollution on human health and the environment. Here are some common technologies used for air pollution control and monitoring:

Air Pollution Control Technologies:

1. Particulate Matter Control: Technologies like electrostatic precipitators (ESPs), fabric filters (baghouses), and cyclones are used to remove particulate matter (PM) from industrial emissions. These devices use electrostatic forces, filtration, or centrifugal force to capture and remove PM from exhaust gases.

2. Flue Gas Desulfurization (FGD): FGD technologies, such as wet scrubbers and dry scrubbers, are used to remove sulfur dioxide (SO_2) from flue gases emitted by power plants and industrial processes. These systems typically use chemicals or sorbents to react with and capture SO_2 before it is released into the atmosphere.

3. Selective Catalytic Reduction (SCR): SCR systems are used to reduce nitrogen oxide (NOx) emissions from power plants and industrial boilers. SCR utilizes a catalyst and a reductant (such as ammonia or urea) to convert NOx into nitrogen and water vapor through a chemical reaction.

4. Volatile Organic Compound (VOC) Control: Technologies

like thermal oxidizers, catalytic oxidizers, and carbon adsorption systems are used to control VOC emissions from industrial processes and storage tanks. These technologies either combust or adsorb VOCs to prevent their release into the atmosphere.

5. Industrial Emission Standards: Governments and regulatory agencies often enforce emission standards and regulations to control and limit air pollutant emissions from various industries. These standards set limits on pollutant concentrations and require industries to install pollution control technologies to meet compliance requirements.

Air Pollution Monitoring Technologies:

1. Ambient Air Quality Monitoring Stations: These stations consist of sensors and instruments that measure the concentration of various air pollutants in the ambient air. They are typically located in urban areas and provide real-time data on air quality for regulatory purposes and public awareness.

2. Remote Sensing and Satellite Monitoring: Remote sensing technologies, including satellite-based monitoring systems, enable the assessment and monitoring of air pollution over large geographic areas. These systems provide valuable data on pollutant distribution, sources, and trends.

3. Mobile Monitoring: Mobile monitoring involves the use of sensors and instruments installed on vehicles to measure air pollution levels while traveling throughout a particular area. This approach helps identify pollution hotspots and assess air quality variations in different locations.

4. Indoor Air Quality (IAQ) Monitoring: IAQ monitoring systems measure pollutants present indoors, such as volatile organic compounds, carbon dioxide, and

particulate matter. These systems help identify potential sources of indoor air pollution and assess the effectiveness of ventilation and filtration systems.

5. Personal Air Quality Monitors: Personal air quality monitors are portable devices that individuals can carry to measure their immediate environment's air pollution levels. These devices provide personal exposure data and can help individuals make informed decisions about their activities and protect their health.

6. Air Quality Modeling: Air quality modeling uses computer simulations to predict and understand pollutant dispersion and concentration patterns. It helps identify pollution sources, assess the effectiveness of control measures, and develop air quality management strategies.

These technologies, combined with effective policies, regulations, and public awareness, are essential for mitigating air pollution, improving air quality, and protecting human health and the environment. By monitoring and controlling air pollution, we can work towards creating cleaner and healthier living environments for present and future generations.

Strategies for minimizing emissions and promoting clean air

Strategies for minimizing emissions and promoting clean air are crucial for mitigating air pollution, improving air quality, and protecting human health and the environment. Here are some key strategies that can be implemented at various levels:

1. Transition to Cleaner Energy Sources: Shifting from fossil fuels to cleaner and renewable energy sources such as solar, wind, and hydroelectric power reduces emissions from power generation. Promoting energy efficiency and incentivizing the use of clean technologies can further reduce greenhouse gas emissions and improve air quality.

2. Improve Industrial Processes: Encouraging industries to adopt cleaner production technologies, such as advanced combustion systems, energy-efficient equipment, and pollution control technologies, can significantly reduce emissions from manufacturing processes. Implementing strict emission standards and providing incentives for adopting cleaner technologies can drive the industry's transition to more sustainable practices.

3. Enhance Transportation Systems: Developing and promoting sustainable transportation solutions can help reduce emissions from the transportation sector, which is a significant contributor to air pollution. Strategies include expanding public transportation networks, promoting electric vehicles and hybrids,

improving fuel efficiency standards, and encouraging active modes of transportation such as walking and cycling.

4. Strengthen Emission Standards and Regulations: Governments and regulatory agencies can set and enforce stringent emission standards for industries, vehicles, and other pollution sources. This includes establishing limits on pollutant emissions, promoting the use of clean technologies, and implementing monitoring and reporting systems to ensure compliance.

5. Promote Sustainable Land Use and Urban Planning: Designing cities and communities with an emphasis on compact and sustainable urban development can reduce transportation-related emissions and promote cleaner air. This involves creating walkable and bike-friendly neighborhoods, preserving green spaces, and integrating public transit systems to reduce the reliance on individual vehicles.

6. Encourage Waste Management Practices: Implementing proper waste management practices, including recycling, composting, and waste-to-energy conversion, reduces the release of pollutants into the air. Effective waste management strategies reduce the need for landfilling, minimize methane emissions from organic waste, and promote resource recovery.

7. Public Awareness and Education: Educating the public about the importance of clean air and the individual actions they can take to minimize emissions is crucial. Public awareness campaigns can encourage behavior changes such as reducing energy consumption, using public transportation, practicing proper waste management, and supporting sustainable initiatives.

8. International Collaboration: Air pollution is a global issue that requires international collaboration. Governments, organizations, and communities can

work together to share best practices, exchange knowledge, and develop joint initiatives to address cross-border air pollution challenges.

By implementing these strategies, it is possible to minimize emissions, improve air quality, and create healthier and more sustainable environments. Collaborative efforts among governments, businesses, communities, and individuals are essential for achieving clean air goals and protecting the well-being of current and future generations.

Urbanization and its environmental impacts

Urbanization, the process of population growth and the migration of people from rural to urban areas, has significant environmental impacts. While urban areas offer various economic and social opportunities, they also pose challenges for sustainability and environmental management. Here are some of the environmental impacts associated with urbanization:

1. Habitat Loss and Fragmentation: Urban development often results in the conversion of natural habitats into built environments. This leads to the loss of biodiversity and fragmentation of ecosystems, disrupting natural habitats and impacting wildlife populations.

2. Increased Energy Consumption: Urban areas tend to have higher energy demands due to factors such as increased population density, increased infrastructure, and higher levels of economic activity. This results in increased greenhouse gas emissions, air pollution, and resource consumption.

3. Air Pollution: Urban areas often experience higher levels of air pollution due to factors such as vehicular emissions, industrial activities, and energy consumption. The release of pollutants such as particulate matter, nitrogen oxides, and volatile organic compounds can have detrimental effects on air quality and human health.

4. Water Resource Stress: Urbanization places significant pressure on water resources. Increased water demand for households, industries, and agriculture can lead to water scarcity and stress on water supply systems.

Urban runoff, contaminated by pollutants from various sources, also contributes to water pollution.

5. Waste Generation: Urban areas generate significant amounts of waste due to increased consumption and population density. Improper waste management can result in pollution of land, water bodies, and air. It poses challenges for waste disposal, recycling, and the management of hazardous waste.

6. Heat Island Effect: Urban areas often experience the urban heat island effect, where built environments absorb and retain heat, leading to higher temperatures compared to surrounding rural areas. This effect is exacerbated by factors such as high-rise buildings, concrete, asphalt, and reduced vegetation cover, impacting local climate patterns and energy demands for cooling.

7. Increased Water Runoff and Flooding: Urbanization alters the natural hydrological cycle, resulting in increased water runoff during rainfall events. This can overwhelm urban drainage systems and lead to urban flooding. Impervious surfaces and inadequate infrastructure for stormwater management exacerbate this issue.

8. Impacts on Public Health: Urbanization can have adverse effects on public health. Poor air quality, water pollution, inadequate sanitation, and overcrowding in urban areas can contribute to the spread of diseases and respiratory illnesses. Limited access to green spaces and recreational areas can also impact mental and physical well-being.

To address these environmental impacts, sustainable urban planning and management practices are essential. This includes incorporating green infrastructure, promoting energy-efficient buildings, improving public transportation systems, implementing waste management strategies, protecting natural areas, and enhancing environmental regulations. By adopting sustainable approaches, urban areas can mitigate their environmental footprint and create more livable, resilient, and environmentally friendly cities.

Sustainable urban planning and design principles

Sustainable urban planning and design principles aim to create cities and communities that are environmentally friendly, socially equitable, and economically viable. These principles guide the development and management of urban areas to minimize negative environmental impacts and promote sustainability. Here are some key sustainable urban planning and design principles:

1. Compact and Mixed-Use Development: Promote compact and mixed-use development to reduce urban sprawl and minimize the need for long commutes. Design neighborhoods that integrate residential, commercial, and recreational spaces, allowing for walkability and reducing the dependence on private vehicles.

2. Smart Growth: Implement smart growth principles that focus on creating sustainable, well-connected, and vibrant communities. Encourage infill development, revitalization of existing areas, and the efficient use of existing infrastructure to minimize resource consumption and preserve natural areas.

3. Sustainable Transportation: Develop transportation systems that prioritize public transit, cycling, and pedestrian infrastructure. Promote the use of electric and hybrid vehicles, carpooling, and ridesharing. Design streets and neighborhoods to be pedestrian-friendly, with well-connected sidewalks, bike lanes, and public transportation routes.

4. Green Building and Infrastructure: Emphasize green building practices that optimize energy efficiency, water conservation, and waste reduction. Encourage the use of renewable energy sources, efficient building materials, green roofs, and rainwater harvesting systems. Incorporate green spaces, parks, and urban forests to improve air quality and provide recreational areas.

5. Urban Heat Island Mitigation: Implement strategies to mitigate the urban heat island effect. Incorporate green roofs, green walls, and high-albedo materials to reduce heat absorption. Promote urban forestry and the use of shade trees to provide natural cooling and improve air quality.

6. Sustainable Water Management: Adopt integrated water management strategies that promote water conservation, stormwater management, and wastewater treatment. Implement measures such as rainwater harvesting, graywater recycling, and permeable pavements to reduce water consumption and minimize water runoff.

7. Social Equity and Inclusive Communities: Ensure that urban planning and design promote social equity and inclusivity. Design accessible infrastructure, public spaces, and buildings for people of all ages, abilities, and socioeconomic backgrounds. Foster social cohesion, community engagement, and affordable housing options.

8. Preservation of Natural Areas: Protect and preserve natural areas, green spaces, and biodiversity within urban environments. Incorporate urban agriculture, community gardens, and green corridors to enhance ecological connectivity and provide opportunities for local food production.

9. Resilience and Climate Adaptation: Design cities and communities to be resilient to climate change impacts. Consider floodplain management, coastal protection,

and strategies to adapt to changing weather patterns. Integrate climate-responsive design elements, such as green infrastructure and building resilience measures.

10. Community Engagement and Participation: Foster community engagement and participatory decision-making processes. Involve residents, stakeholders, and local communities in the planning and design of their neighborhoods to ensure their needs, values, and perspectives are incorporated.

By applying these sustainable urban planning and design principles, cities can create healthier, more livable, and sustainable environments that benefit both present and future generations. The integration of environmental, social, and economic considerations helps create vibrant and resilient cities that enhance the quality of life for their residents while minimizing their ecological footprint.

Smart cities and innovative urban solutions

Smart cities and innovative urban solutions leverage technology and data to improve the efficiency, sustainability, and quality of life in urban areas. These solutions integrate digital infrastructure, connectivity, and information systems to enhance various aspects of urban living, including transportation, energy, waste management, public safety, and governance. Here are some key components and examples of smart cities and innovative urban solutions:

1. Intelligent Transportation Systems: Smart cities utilize technologies such as real-time traffic monitoring, smart parking systems, and intelligent transportation networks to optimize traffic flow, reduce congestion, and enhance public transportation services. Examples include smart traffic lights that adjust timing based on traffic conditions and mobile apps that provide real-time transit information.

2. Energy Management and Efficiency: Smart cities focus on energy management and efficiency through the integration of renewable energy sources, smart grids, and energy monitoring systems. This includes initiatives like smart meters that enable users to monitor and optimize their energy consumption and the implementation of energy storage systems.

3. Internet of Things (IoT) and Sensor Networks: IoT devices and sensor networks play a crucial role in smart cities by collecting and analyzing data from various urban systems. These systems can include sensors for air quality monitoring, waste management, and water

management. The data collected helps inform decision-making processes and enables better resource allocation and optimization.

4. Sustainable Infrastructure: Smart cities prioritize sustainable infrastructure by incorporating green building practices, renewable energy generation, and efficient water management systems. This can include features like energy-efficient buildings, green roofs, smart irrigation systems, and rainwater harvesting.

5. Citizen Engagement and Participation: Smart cities promote citizen engagement and participation through digital platforms and mobile applications. These platforms allow residents to provide feedback, report issues, access city services, and participate in decision-making processes. Examples include apps for reporting infrastructure problems or crowdsourcing data on environmental conditions.

6. Big Data and Analytics: Smart cities leverage big data and advanced analytics to gain insights and make informed decisions. Data collected from various sources, such as sensors, social media, and administrative systems, can be analyzed to identify patterns, optimize resource allocation, and improve service delivery.

7. Public Safety and Emergency Management: Smart city solutions enhance public safety and emergency management through technologies such as video surveillance, intelligent emergency response systems, and real-time situational awareness platforms. These technologies improve incident response times, enhance public safety measures, and enable effective disaster management.

8. Smart Governance and Open Data: Smart cities promote transparency, efficiency, and accountability through digital governance systems and open data initiatives. This includes online platforms for public services,

open data portals, and digital platforms for citizen engagement in decision-making processes.

9. Waste Management and Recycling: Smart cities implement innovative waste management and recycling solutions, such as smart bins with fill-level sensors, waste collection optimization algorithms, and recycling incentives. These technologies help reduce waste generation, improve collection efficiency, and promote recycling practices.

10. Sustainable Mobility: Smart cities focus on sustainable mobility solutions by integrating electric vehicles, bike-sharing systems, and smart mobility platforms. This includes initiatives like electric vehicle charging infrastructure, bike-sharing apps, and integrated multimodal transportation planning.

By integrating these smart city solutions and innovative urban technologies, cities can improve resource efficiency, enhance quality of life, and promote sustainability. The use of data-driven decision-making and technology-driven solutions helps address urban challenges, optimize resource management, and create more livable and resilient cities for the future.

Importance of transitioning to renewable energy sources

Transitioning to renewable energy sources is of paramount importance for several compelling reasons:

1. Climate Change Mitigation: Renewable energy sources such as solar, wind, hydro, and geothermal power generate electricity with significantly lower greenhouse gas emissions compared to fossil fuels. By shifting away from fossil fuels and embracing renewables, we can reduce the carbon footprint of our energy sector, which is a primary contributor to climate change. This transition is crucial to mitigating the impacts of climate change, such as rising temperatures, extreme weather events, and sea-level rise.

2. Energy Security and Independence: Dependence on finite fossil fuel resources leaves countries vulnerable to price volatility, geopolitical tensions, and supply disruptions. Transitioning to renewable energy sources enhances energy security by diversifying the energy mix and reducing reliance on imported fossil fuels. Countries can tap into their domestic renewable resources, promoting energy independence and resilience.

3. Improved Air Quality and Public Health: Fossil fuel combustion for energy generation releases pollutants such as sulfur dioxide, nitrogen oxides, particulate matter, and volatile organic compounds. These pollutants contribute to air pollution, leading

to respiratory and cardiovascular diseases, premature deaths, and other adverse health effects. Shifting to renewable energy sources significantly reduces harmful emissions, improving air quality and public health outcomes.

4. Job Creation and Economic Growth: The transition to renewable energy sources presents significant economic opportunities. It stimulates the growth of renewable energy industries, creating jobs in manufacturing, installation, operation, and maintenance of renewable energy infrastructure. Investments in renewable energy can drive economic growth, innovation, and technological advancements, fostering a clean energy economy.

5. Sustainable Development and Environmental Stewardship: Embracing renewable energy aligns with the principles of sustainable development and environmental stewardship. Renewable energy sources are inherently sustainable, as they are naturally replenished and do not deplete finite resources. They have a lower environmental impact throughout their lifecycle, from extraction to disposal, compared to fossil fuel-based energy sources.

6. Technological Advancements and Cost Competitiveness: The rapid advancement of renewable energy technologies, coupled with economies of scale, has led to significant cost reductions. Renewable energy sources, such as solar and wind power, have become increasingly competitive with fossil fuels in terms of cost. Continued technological advancements and innovation are expected to further drive down costs and improve the efficiency of renewable energy systems.

7. Global Leadership and International Cooperation: Transitioning to renewable energy sources allows countries to showcase their commitment to environmental sustainability and climate action. By

leading the way in renewable energy deployment, countries can influence global efforts to combat climate change, promote international cooperation, and inspire others to follow suit.

Transitioning to renewable energy sources is essential for a sustainable and resilient future. It is a key strategy to address climate change, improve air quality, enhance energy security, create jobs, and promote sustainable development. Governments, businesses, and individuals all have a role to play in accelerating the transition to renewable energy and reaping the benefits it offers.

Overview of different renewable energy technologies

There are several different renewable energy technologies available today that harness naturally replenished sources of energy. Here is an overview of some of the most common renewable energy technologies:

1. Solar Energy: Solar energy harnesses the power of the sun to generate electricity or heat. Photovoltaic (PV) cells convert sunlight directly into electricity, while solar thermal systems use sunlight to heat water or air for various applications.

2. Wind Energy: Wind energy utilizes the kinetic energy of the wind to generate electricity. Wind turbines capture the wind's energy and convert it into electrical power. Onshore and offshore wind farms are common installations for large-scale wind energy generation.

3. Hydropower: Hydropower harnesses the energy of moving water, such as rivers or tidal currents, to generate electricity. Large-scale hydropower plants use dams and turbines, while smaller-scale installations include run-of-river hydroelectric systems and tidal energy converters.

4. Geothermal Energy: Geothermal energy utilizes the heat from the Earth's core to generate electricity or provide heating and cooling. Geothermal power plants extract heat from underground reservoirs of hot water or steam and convert it into electricity.

5. Biomass Energy: Biomass energy involves using organic

materials, such as wood, crops, agricultural residues, and organic waste, to generate heat or electricity. Biomass can be burned directly, converted into biogas through anaerobic digestion, or used to produce biofuels.

6. Ocean Energy: Ocean energy technologies capture the power of tides, waves, and ocean currents to generate electricity. Tidal energy systems use the rise and fall of tides, wave energy converters capture the energy from waves, and marine current turbines harness the power of ocean currents.

7. Hydrogen Fuel Cells: Hydrogen fuel cells produce electricity by combining hydrogen and oxygen, with the only byproduct being water vapor. Hydrogen can be derived from various renewable sources, such as electrolysis of water using renewable electricity, and used in fuel cells to power vehicles or generate electricity.

8. Concentrated Solar Power (CSP): CSP systems use mirrors or lenses to concentrate sunlight onto a receiver, generating high-temperature heat that is used to produce steam and drive turbines for electricity generation.

These renewable energy technologies offer a diverse range of options for generating clean and sustainable energy. They contribute to reducing greenhouse gas emissions, improving energy security, and promoting sustainable development. Each technology has its own specific characteristics, advantages, and deployment considerations, and their suitability depends on factors such as resource availability, geographic location, and local conditions. Combining multiple renewable energy technologies in an integrated energy system can maximize the benefits and enhance overall energy sustainability.

Strategies for improving energy efficiency in various sectors

Improving energy efficiency is a crucial strategy for reducing energy consumption, lowering greenhouse gas emissions, and promoting sustainable development. Here are some strategies for improving energy efficiency in various sectors:

1. Buildings:
   - Enhance insulation: Improve building envelope insulation to reduce heat transfer and minimize energy loss.
   - Energy-efficient lighting: Replace traditional lighting systems with energy-efficient LED or CFL bulbs.
   - Efficient HVAC systems: Install high-efficiency heating, ventilation, and air conditioning (HVAC) systems and ensure regular maintenance.
   - Smart controls: Use smart thermostats, occupancy sensors, and programmable controls to optimize energy use.
   - Building codes and standards: Implement and enforce energy codes and standards for new construction and retrofits.

2. Transportation:
   - Promote public transit: Improve and expand public transportation networks to encourage the use of mass transit systems.
   - Encourage active transportation: Develop

infrastructure for walking, cycling, and other non-motorized modes of transportation.
- Efficient vehicles: Promote the use of fuel-efficient and low-emission vehicles, such as hybrids and electric vehicles.
- Carpooling and ridesharing: Encourage carpooling and ridesharing programs to reduce the number of vehicles on the road.
- Eco-driving: Educate drivers about fuel-efficient driving techniques, such as avoiding rapid acceleration and maintaining proper tire pressure.

3. Industrial Sector:
- Energy audits: Conduct regular energy audits to identify areas for energy efficiency improvements.
- Process optimization: Implement process optimization techniques to minimize energy waste and optimize equipment operation.
- Upgraded equipment: Upgrade machinery and equipment to more energy-efficient models.
- Waste heat recovery: Capture and utilize waste heat generated during industrial processes.
- Employee engagement: Train and engage employees in energy-saving practices and encourage their participation in energy efficiency initiatives.

4. Agriculture:
- Efficient irrigation: Use precision irrigation techniques and technologies to optimize water and energy use in agriculture.
- Energy-efficient equipment: Utilize energy-efficient machinery, such as pumps and irrigation systems.
- Renewable energy integration: Explore the use of renewable energy systems, such as solar or

wind power, to meet agricultural energy needs.

- Sustainable practices: Implement sustainable farming practices that minimize energy-intensive activities and optimize resource use.

5. Information Technology:
   - Data center efficiency: Improve data center cooling and ventilation systems to reduce energy consumption.
   - Virtualization and cloud computing: Consolidate servers and utilize virtualization and cloud computing to reduce energy demand.
   - Energy-efficient equipment: Use energy-efficient computers, servers, and networking equipment.
   - Power management: Implement power management features to optimize energy use in IT equipment.

6. Public Sector and Institutions:
   - Energy management systems: Deploy energy management systems to monitor and control energy use in public buildings and facilities.
   - Energy awareness programs: Raise awareness among employees and stakeholders about energy conservation practices.
   - Energy performance contracts: Enter into energy performance contracts to incentivize energy efficiency improvements in public buildings.
   - Retrofit programs: Implement retrofit programs to upgrade public buildings with energy-efficient technologies and systems.

These strategies, when implemented comprehensively and supported by policies and incentives, can significantly improve energy efficiency and contribute to a more sustainable and low-carbon future. Collaboration between governments, businesses, and individuals is essential for achieving meaningful energy efficiency gains across all sectors of the economy.

Understanding the impacts of climate change

Climate change refers to long-term shifts in temperature patterns and weather conditions on Earth. It is primarily caused by human activities that release greenhouse gases (GHGs) into the atmosphere, such as burning fossil fuels, deforestation, and industrial processes. The impacts of climate change are far-reaching and affect various aspects of the natural environment, ecosystems, and human societies. Here are some key impacts of climate change:

1. Rising Temperatures: Global average temperatures have been steadily increasing, resulting in more frequent and intense heatwaves. Higher temperatures can lead to heat-related illnesses and deaths, affect crop yields, disrupt ecosystems, and exacerbate drought conditions.

2. Extreme Weather Events: Climate change contributes to more frequent and severe extreme weather events, including hurricanes, cyclones, floods, droughts, and wildfires. These events can cause significant damage to infrastructure, homes, and livelihoods, and lead to loss of life.

3. Sea Level Rise: As global temperatures rise, glaciers and ice caps melt, contributing to the rise in sea levels. Rising sea levels pose a threat to coastal communities and low-lying areas, increasing the risk of coastal erosion, inundation, and saltwater intrusion into freshwater resources.

4. Changes in Precipitation Patterns: Climate change affects precipitation patterns, leading to shifts in rainfall distribution and intensity. Some regions may

experience increased rainfall and flooding, while others may face reduced rainfall and drought conditions. These changes impact agriculture, water availability, and the overall ecosystem balance.

5. Ecosystem Disruption: Climate change disrupts ecosystems and biodiversity. Species may struggle to adapt to rapidly changing conditions, leading to shifts in migration patterns, altered habitats, and potential extinction risks. Loss of biodiversity can have cascading effects on ecosystem functioning and services, including pollination, nutrient cycling, and disease regulation.

6. Agriculture and Food Security: Changes in temperature, rainfall, and growing seasons affect agricultural productivity and food security. Crop yields can decline due to heat stress, droughts, floods, and changes in pest and disease dynamics. Climate change poses risks to food production, distribution, and affordability, impacting global food security.

7. Human Health: Climate change influences the spread of infectious diseases, affects air quality, and exacerbates health risks. Heatwaves can lead to heat-related illnesses and fatalities, while changing precipitation patterns can influence the prevalence of waterborne diseases. Additionally, climate change can contribute to air pollution, allergies, and respiratory illnesses.

8. Economic Impacts: Climate change has significant economic implications. The costs associated with extreme weather events, infrastructure damage, health impacts, and disruptions to agriculture and industry can be substantial. Businesses and economies may face challenges, particularly in sectors dependent on climate-sensitive resources.

It is crucial to address climate change through mitigation and adaptation strategies. Mitigation involves reducing greenhouse gas emissions through measures like transitioning to renewable energy sources, energy efficiency improvements, and sustainable land use practices. Adaptation involves preparing for and managing the impacts of climate change, such as implementing climate-resilient infrastructure, enhancing water management, and developing early warning systems. Effective climate action requires international cooperation, policy frameworks, technological advancements, and individual behavioral changes to create a sustainable and resilient future.

Strategies for adapting to climate change effects

Adapting to the effects of climate change is crucial for building resilience and reducing vulnerability to its impacts. Adaptation strategies involve preparing for and responding to changes in climate conditions to protect communities, ecosystems, and economies. Here are some key strategies for adapting to climate change effects:

1. Climate Risk Assessment: Conduct comprehensive climate risk assessments to identify the specific climate-related hazards and vulnerabilities in a particular region or sector. This assessment helps in understanding the potential impacts and planning appropriate adaptation measures.

2. Water Management: Develop robust water management strategies to address changing precipitation patterns and water availability. This can include implementing water conservation measures, enhancing water storage and infrastructure, promoting efficient irrigation techniques, and managing water allocation in a changing climate.

3. Coastal Protection: Implement measures to protect coastal areas from sea-level rise and increased storm surge risks. This can involve building or enhancing coastal defenses, restoring natural ecosystems like mangroves and coral reefs, and implementing coastal zoning and land use planning.

4. Enhanced Infrastructure Design: Incorporate climate

resilience into the design and construction of infrastructure systems. This includes considering the impacts of climate change in infrastructure planning, building more resilient buildings and transportation networks, and adopting climate-smart engineering techniques.

5. Agriculture and Food Security: Promote climate-smart agriculture practices that enhance resilience to climate change. This can include diversifying crop varieties, adopting conservation agriculture techniques, improving water management in agriculture, and implementing early warning systems for pests and diseases.

6. Ecosystem Restoration and Conservation: Protect and restore natural ecosystems, such as forests, wetlands, and coastal habitats, as they provide essential ecosystem services and play a crucial role in climate regulation. Conservation efforts can enhance biodiversity, carbon sequestration, and water resource management.

7. Health Systems Strengthening: Strengthen healthcare systems to cope with the health impacts of climate change. This can involve improving disease surveillance and response systems, developing heat action plans, and enhancing public health education and awareness.

8. Community Engagement and Capacity Building: Engage and involve local communities in adaptation planning and decision-making processes. Build community resilience through education, training, and capacity building initiatives that empower individuals and communities to adapt to climate change.

9. Financial Mechanisms: Establish and enhance financial mechanisms to support adaptation efforts. This can include creating dedicated funds for climate adaptation, providing incentives and financial support for climate-resilient projects, and promoting risk-transfer

     mechanisms such as insurance and climate risk pooling.

10. Knowledge Sharing and Collaboration: Foster knowledge sharing, collaboration, and information exchange among stakeholders, including governments, researchers, communities, and non-governmental organizations. Sharing best practices, lessons learned, and scientific findings helps to accelerate adaptation efforts.

Adaptation strategies should be context-specific, taking into account the unique characteristics, vulnerabilities, and opportunities of each region or sector. It is essential to integrate adaptation into policies, plans, and decision-making processes at all levels to ensure long-term resilience and sustainability in the face of climate change.

Mitigation measures to reduce greenhouse gas emissions

Mitigation measures are essential to reduce greenhouse gas (GHG) emissions and mitigate climate change. These measures aim to decrease the release of GHGs into the atmosphere and promote sustainable practices. Here are some key mitigation measures:

1. Transition to Renewable Energy: Increase the use of renewable energy sources such as solar, wind, hydro, and geothermal power. This involves transitioning away from fossil fuel-based energy generation and promoting the deployment of clean energy technologies.

2. Energy Efficiency Improvements: Enhance energy efficiency across various sectors, including buildings, transportation, and industry. This includes implementing energy-efficient technologies, improving insulation, adopting efficient lighting and appliances, and optimizing industrial processes.

3. Sustainable Transportation: Promote low-carbon transportation options such as public transit, cycling, walking, and electric vehicles. Encourage the development of charging infrastructure for electric vehicles, promote carpooling and ridesharing, and invest in public transportation systems.

4. Sustainable Land Use and Forestry: Protect and restore forests, as they act as carbon sinks and absorb significant amounts of CO_2. Implement sustainable land management practices, reduce deforestation, and promote afforestation and reforestation efforts.

5. Energy-Efficient Buildings: Improve energy efficiency in buildings through measures like energy-efficient building design, insulation, efficient heating and cooling systems, and smart building technologies. Enhance energy performance standards for new constructions and encourage retrofits of existing buildings.

6. Circular Economy and Waste Management: Promote waste reduction, recycling, and proper waste management practices. Implement recycling programs, encourage waste reduction initiatives, and support the development of circular economy models that reduce resource consumption and waste generation.

7. Sustainable Agriculture and Food Systems: Promote sustainable agricultural practices that reduce GHG emissions, enhance carbon sequestration, and improve resilience. This includes practices like agroforestry, conservation agriculture, organic farming, and improved livestock management.

8. Industry and Manufacturing: Implement cleaner production techniques and technologies in industries to reduce GHG emissions. Improve energy efficiency in manufacturing processes, adopt renewable energy sources, and implement carbon capture and storage technologies where applicable.

9. Carbon Pricing and Financial Mechanisms: Implement carbon pricing mechanisms such as carbon taxes or emissions trading systems to incentivize emission reductions. Promote green finance, investments in low-carbon technologies, and provide financial incentives for sustainable practices.

10. Education and Awareness: Raise awareness about the importance of mitigating climate change and promoting sustainable practices. Education and public awareness campaigns can encourage individuals, communities, and businesses to adopt climate-friendly

behaviors and support mitigation efforts.

These mitigation measures should be implemented at various levels, including international, national, regional, and local levels, and require collaboration and commitment from governments, businesses, communities, and individuals. Combining multiple measures and approaches can significantly contribute to reducing GHG emissions and mitigating the impacts of climate change.

Role of government policies and regulations

Government policies and regulations play a crucial role in addressing climate change and facilitating the transition to a low-carbon economy. Here are some key ways in which government policies and regulations can drive climate action:

1. Setting Emission Reduction Targets: Governments can establish ambitious emission reduction targets that align with international climate agreements and scientific recommendations. These targets provide a clear direction for climate action and help guide policy development and implementation.

2. Implementing Carbon Pricing: Governments can introduce carbon pricing mechanisms, such as carbon taxes or emissions trading systems, to put a price on carbon emissions. Carbon pricing incentivizes businesses and individuals to reduce their emissions and invest in low-carbon alternatives.

3. Renewable Energy Policies: Governments can establish supportive policies and regulations to promote the development and deployment of renewable energy sources. This includes providing incentives, subsidies, and feed-in tariffs for renewable energy projects, as well as streamlining the permitting and regulatory processes.

4. Energy Efficiency Standards: Governments can set energy efficiency standards for buildings, appliances, and vehicles. These standards ensure that energy-consuming products meet certain efficiency requirements, encouraging the adoption of energy-

efficient technologies and reducing overall energy consumption.

5. Support for Research and Development: Governments can invest in research and development (R&D) to advance low-carbon technologies and solutions. Funding R&D initiatives can accelerate innovation, improve the cost-effectiveness of clean technologies, and drive their market deployment.

6. Sustainable Transport Policies: Governments can implement policies to promote sustainable transportation options, such as public transit, cycling infrastructure, and electric vehicles. This includes providing incentives for electric vehicle adoption, expanding public transportation networks, and implementing vehicle emission standards.

7. Land Use and Forest Protection: Governments can develop policies and regulations to protect forests, promote sustainable land use practices, and discourage deforestation. This includes measures such as land-use planning, reforestation programs, and regulations against unsustainable agricultural practices.

8. Building Codes and Standards: Governments can establish and enforce building codes and standards that require energy-efficient designs, materials, and systems for new constructions and retrofits. This encourages the adoption of sustainable building practices and improves energy efficiency in the construction sector.

9. International Cooperation and Agreements: Governments play a crucial role in international climate negotiations and cooperation. They can work collaboratively to develop and implement global agreements, such as the Paris Agreement, and engage in efforts to address climate change at a global scale.

10. Education and Awareness: Governments can invest in public education and awareness campaigns to promote climate literacy and encourage behavioral changes. This

can include programs that highlight the benefits of sustainable practices, provide information on energy conservation, and promote climate-friendly behaviors.

Government policies and regulations provide the necessary framework and incentives to drive climate action, spur innovation, and create a conducive environment for businesses, communities, and individuals to adopt sustainable practices. By setting clear targets, implementing effective regulations, and providing supportive measures, governments can catalyze the transition to a low-carbon, climate-resilient future.

International environmental agreements and frameworks

International environmental agreements and frameworks are crucial for addressing global environmental challenges and promoting international cooperation on sustainability issues. These agreements provide a platform for countries to come together, set common goals, share best practices, and work collectively towards protecting the environment. Here are some key international environmental agreements and frameworks:

1. United Nations Framework Convention on Climate Change (UNFCCC): The UNFCCC is an international treaty adopted in 1992 with the goal of stabilizing greenhouse gas concentrations in the atmosphere and preventing dangerous anthropogenic interference with the climate system. It provides the framework for international cooperation on climate change, including the annual Conference of the Parties (COP) meetings and the negotiation of specific agreements such as the Kyoto Protocol and the Paris Agreement.

2. Paris Agreement: The Paris Agreement, adopted in 2015 under the UNFCCC, is a landmark international treaty aimed at limiting global warming to well below 2 degrees Celsius above pre-industrial levels and pursuing efforts to limit the temperature increase to 1.5 degrees Celsius. It establishes binding commitments for countries to reduce their greenhouse gas emissions, adapt to the impacts of climate change, and provide financial and technological support to developing

countries.

3. Convention on Biological Diversity (CBD): The CBD is an international treaty adopted in 1992 with the objective of conserving biodiversity, promoting sustainable use of natural resources, and ensuring the fair and equitable sharing of benefits arising from genetic resources. The CBD provides a framework for countries to develop national biodiversity strategies, protect ecosystems, and conserve species diversity.

4. Montreal Protocol on Substances that Deplete the Ozone Layer: The Montreal Protocol, adopted in 1987, is an international treaty designed to protect the ozone layer by phasing out the production and consumption of ozone-depleting substances (ODSs), such as chlorofluorocarbons (CFCs) and hydrochlorofluorocarbons (HCFCs). The agreement has been highly successful in reducing the production and use of ODSs and has contributed to the healing of the ozone layer.

5. Sustainable Development Goals (SDGs): The SDGs are a set of 17 global goals adopted by the United Nations in 2015, aimed at addressing various social, economic, and environmental challenges. The goals include targets related to poverty eradication, education, gender equality, clean energy, climate action, and the protection of ecosystems. The SDGs provide a comprehensive framework for sustainable development efforts, with a target deadline of 2030.

6. Basel Convention on the Control of Transboundary Movements of Hazardous Wastes and Their Disposal: The Basel Convention, adopted in 1989, is an international treaty designed to regulate the transboundary movement and disposal of hazardous wastes. It aims to minimize the generation of hazardous waste, ensure the environmentally sound management of such waste, and prevent illegal dumping or export of

hazardous materials to developing countries.

7. Ramsar Convention on Wetlands: The Ramsar Convention, adopted in 1971, is an international treaty for the conservation and wise use of wetlands. It promotes the conservation of wetland ecosystems, including lakes, rivers, marshes, and coastal areas, and encourages their sustainable use for the benefit of present and future generations.

These international agreements and frameworks provide a platform for countries to collaborate, exchange information, and coordinate actions to address pressing environmental challenges. They foster global cooperation, knowledge sharing, and policy development, helping to create a more sustainable and resilient world. Implementation and compliance with these agreements are critical for achieving global environmental goals and ensuring the well-being of both current and future generations.

Community engagement and public participation

Community engagement and public participation are vital components of effective environmental management and decision-making processes. Engaging communities and involving the public in environmental initiatives allows for their perspectives, knowledge, and concerns to be considered, ultimately leading to more inclusive and sustainable outcomes. Here are some key aspects and benefits of community engagement and public participation:

1. Access to Local Knowledge: Communities possess valuable local knowledge about their environment, including traditional practices, ecological insights, and understanding of local resources. Engaging with communities allows for the exchange of this knowledge, which can enhance the effectiveness and relevance of environmental initiatives.

2. Improved Decision-Making: Public participation enables stakeholders to have a voice in decision-making processes related to environmental issues. It ensures that a range of perspectives and interests are taken into account, leading to more informed, balanced, and robust decision-making outcomes.

3. Building Trust and Collaboration: Community engagement fosters trust and collaboration between stakeholders, including community members, government agencies, non-governmental organizations, and businesses. It creates opportunities

for open dialogue, mutual understanding, and joint problem-solving, leading to stronger relationships and more effective partnerships.

4. Increased Ownership and Responsibility: Involving the community in environmental initiatives fosters a sense of ownership and responsibility for the outcomes. When people feel engaged and included in decision-making processes, they are more likely to support and actively participate in the implementation of environmental projects and initiatives.

5. Enhanced Environmental Education and Awareness: Community engagement provides a platform for environmental education, raising awareness about the importance of environmental issues, and promoting sustainable practices. It allows for the sharing of information, knowledge, and best practices, empowering individuals and communities to make informed choices and take actions to protect their environment.

6. Social and Economic Benefits: Engaging communities in environmental initiatives can lead to social and economic benefits. It can create opportunities for local employment, capacity building, and the development of sustainable livelihoods. Community-based initiatives also have the potential to enhance social cohesion and improve the well-being of community members.

7. Conflict Resolution and Consensus-Building: Community engagement processes facilitate conflict resolution by providing a forum for different perspectives to be heard and understood. Through open dialogue and active participation, stakeholders can work towards consensus, finding common ground and shared solutions to complex environmental challenges.

8. Long-Term Sustainability: Engaging communities ensures that environmental initiatives are aligned with local needs, values, and aspirations. By involving

the public in the planning, implementation, and monitoring of projects, the long-term sustainability and success of these initiatives are more likely to be achieved.

Community engagement and public participation are essential for effective environmental management, as they promote transparency, inclusivity, and accountability. They enable stakeholders to work together towards shared environmental goals and contribute to the overall well-being of communities and ecosystems.

Showcase of real-world environmental engineering projects

There are numerous real-world environmental engineering projects that showcase innovative solutions to environmental challenges. Here are a few examples:

1. Solar Power Plants: Large-scale solar power plants, such as the Noor Complex in Morocco and the Kamuthi Solar Power Project in India, demonstrate the potential of solar energy to provide clean and renewable electricity on a significant scale. These projects utilize photovoltaic technology to harness solar energy and contribute to reducing greenhouse gas emissions.

2. Wastewater Treatment Plants: Advanced wastewater treatment plants employ cutting-edge technologies to treat and purify wastewater before it is released into the environment. Examples include the Western Corridor Recycled Water Scheme in Australia and the Dubai Wastewater Treatment Plant in the United Arab Emirates. These projects ensure that wastewater is properly treated, reducing water pollution and promoting the sustainable use of water resources.

3. Sustainable Urban Planning: Cities around the world are implementing sustainable urban planning initiatives to create more livable and environmentally friendly urban environments. The Vauban district in Freiburg, Germany, is a renowned example of sustainable urban development with its focus on energy-efficient buildings, pedestrian-friendly design, and extensive use

of renewable energy sources.

4. Green Building Design: Green building projects incorporate environmentally friendly design principles and technologies to minimize resource consumption and reduce the environmental impact of buildings. The Edge Building in Amsterdam, Netherlands, is one such example. It is one of the most sustainable office buildings in the world, featuring energy-efficient systems, smart technologies, and innovative design elements.

5. Waste Management and Recycling Facilities: State-of-the-art waste management and recycling facilities help divert waste from landfills and promote recycling. The Puente Hills Materials Recovery Facility in California, USA, is an advanced recycling facility that processes and recycles a significant portion of the region's waste, contributing to waste reduction and resource conservation.

6. Renewable Energy Integration on Islands: Islands face unique energy challenges due to their isolation and limited access to traditional energy sources. Projects like the El Hierro Renewable Energy Project in the Canary Islands, Spain, demonstrate successful integration of renewable energy systems, including wind and hydro power, to achieve energy self-sufficiency and reduce reliance on fossil fuels.

7. Climate Resilience Infrastructure: Climate-resilient infrastructure projects are designed to withstand the impacts of climate change, such as increased flooding and extreme weather events. The 3D-Printed Bridge in the Netherlands showcases innovative construction techniques and materials that are both resilient and sustainable, contributing to climate adaptation efforts.

These are just a few examples of real-world environmental engineering projects that address various environmental challenges. Each project highlights the application of engineering principles, innovative technologies, and sustainable practices to create positive environmental impacts and contribute to a more sustainable future.

Examining their impacts and lessons learned

Examining the impacts and lessons learned from real-world environmental engineering projects can provide valuable insights into their effectiveness and potential for replication and scalability. Here are some common impacts and lessons learned from such projects:

1. Environmental Impact: Real-world environmental engineering projects often have direct positive impacts on the environment. For example, solar power plants reduce greenhouse gas emissions by displacing fossil fuel-based electricity generation. Wastewater treatment plants improve water quality and protect aquatic ecosystems. Waste management and recycling facilities reduce landfill waste and promote resource conservation. Assessing and quantifying these environmental impacts helps evaluate the success of projects and informs future decision-making.

2. Social and Economic Benefits: Environmental engineering projects can bring significant social and economic benefits to communities. They often create jobs during construction and operation phases, stimulate local economies, and improve the quality of life for residents. For example, sustainable urban planning projects can enhance livability, promote walkability, and improve public health. Green building projects can provide healthier indoor environments, reduce energy costs, and enhance occupant comfort.

3. Technology Innovation: Real-world projects often drive technological advancements in environmental

engineering. They push the boundaries of what is possible and encourage the development and deployment of new and improved technologies. For instance, the integration of renewable energy on islands has led to innovations in energy storage systems, grid management, and hybrid energy solutions. These technological advancements have broader implications beyond the specific project and can be applied to other contexts.

4. Lessons on Collaboration and Stakeholder Engagement: Environmental engineering projects require collaboration among various stakeholders, including government agencies, local communities, non-governmental organizations, and private sector entities. Lessons learned from successful projects highlight the importance of effective stakeholder engagement, clear communication, and inclusive decision-making processes. Building trust and engaging stakeholders early on can lead to better project outcomes and community acceptance.

5. Policy and Regulatory Implications: Real-world projects often have policy and regulatory implications, highlighting the need for supportive frameworks. Successful projects can inform policy development and encourage the adoption of regulations that promote environmental sustainability. For example, the success of renewable energy projects can lead to the implementation of supportive policies, such as feed-in tariffs or renewable portfolio standards.

6. Scalability and Replicability: Lessons learned from real-world projects can help identify factors that contribute to their scalability and replicability. Understanding the key drivers of success and challenges faced during implementation can guide future projects and facilitate their expansion to other locations or contexts. Lessons related to project financing, community engagement

strategies, and technological feasibility can inform the replication of successful initiatives.

7. Continuous Monitoring and Evaluation: Monitoring and evaluating the long-term impacts of environmental engineering projects is crucial. It helps assess their effectiveness, identify areas for improvement, and ensure accountability. Robust monitoring and evaluation mechanisms provide feedback on project performance, help refine strategies, and support evidence-based decision-making.

Examining the impacts and lessons learned from real-world environmental engineering projects is essential for advancing sustainable practices and shaping future initiatives. It allows for the identification of best practices, the avoidance of pitfalls, and the replication of successful models to tackle environmental challenges effectively.

Highlighting success stories and best practices

Success stories and best practices in environmental engineering demonstrate effective approaches and strategies in addressing environmental challenges. These examples inspire and provide valuable lessons for future projects. Here are a few success stories and best practices:

1. Masdar City, United Arab Emirates: Masdar City is a sustainable urban development project in Abu Dhabi that showcases innovative technologies and practices. The city aims to be carbon-neutral and zero waste, with a focus on renewable energy, energy-efficient buildings, and smart transportation systems. It demonstrates the integration of sustainable technologies and urban planning principles, serving as a model for future sustainable cities.

2. The Three Gorges Dam, China: The Three Gorges Dam on the Yangtze River is the world's largest hydroelectric power station. It generates clean energy, reduces greenhouse gas emissions, and provides flood control benefits. The project incorporates innovative engineering solutions and environmental management practices to minimize ecological impacts and ensure sustainable water management.

3. The High Line, United States: The High Line is an elevated park built on a disused railway track in New York City. It showcases the transformation of urban infrastructure into a green public space. The

project revitalizes the area, enhances biodiversity, and promotes sustainable urban design. It serves as an inspiration for converting underutilized spaces into sustainable and vibrant community assets.

4. The Great Green Wall Initiative, Africa: The Great Green Wall is an ambitious project aimed at combatting desertification and land degradation in the Sahel region of Africa. It involves planting a wall of trees across the continent to restore degraded land, combat climate change, and improve livelihoods. The project showcases the importance of nature-based solutions in addressing environmental challenges and fostering sustainable development.

5. The Curitiba Bus Rapid Transit (BRT) System, Brazil: Curitiba's BRT system is a renowned example of efficient and sustainable public transportation. It integrates dedicated bus lanes, smart traffic management, and well-designed stations to provide affordable and reliable transportation. The BRT system reduces traffic congestion, improves air quality, and promotes sustainable urban mobility.

6. The Kielder Water Scheme, United Kingdom: The Kielder Water Scheme is a water management project in Northumberland, UK, that supplies water to the region while protecting ecosystems. The project demonstrates sustainable water resource management practices, including reservoir management, water conservation measures, and ecosystem restoration. It showcases the importance of balancing water needs with environmental protection.

7. The European Union Emissions Trading System (EU ETS): The EU ETS is the world's largest carbon market, aiming to reduce greenhouse gas emissions in Europe. It sets a cap on emissions from industrial sectors and establishes a trading system for carbon allowances. The system incentivizes emission reductions, encourages

investments in low-carbon technologies, and showcases the potential of market-based mechanisms to mitigate climate change.

These success stories and best practices highlight innovative approaches, technological advancements, and sustainable management strategies. They demonstrate the feasibility and benefits of implementing environmentally friendly solutions, providing inspiration and guidance for future environmental engineering projects worldwide.

Emerging trends in environmental engineering

Environmental engineering is a field that constantly evolves to address new challenges and leverage technological advancements. Here are some emerging trends in environmental engineering:

1. Climate Resilience and Adaptation: With the increasing impacts of climate change, there is a growing focus on developing climate-resilient infrastructure and implementing adaptation strategies. Environmental engineers are involved in designing and implementing projects that enhance resilience to extreme weather events, sea-level rise, and other climate-related risks.

2. Circular Economy and Resource Management: The shift towards a circular economy, which emphasizes minimizing waste and maximizing resource efficiency, is gaining momentum. Environmental engineers are involved in developing strategies and technologies for waste reduction, recycling, and the recovery of valuable resources from waste streams.

3. Sustainable and Smart Cities: The concept of sustainable and smart cities is gaining prominence as urbanization accelerates. Environmental engineers play a vital role in designing and implementing sustainable urban infrastructure, including efficient transportation systems, energy-efficient buildings, smart water and waste management, and integration of renewable energy sources.

4. Environmental Data Analytics and Modeling: The increasing availability of big data and

advancements in analytics and modeling techniques enable environmental engineers to analyze complex environmental systems and make informed decisions. Data-driven approaches are being used to monitor and assess environmental impacts, optimize resource management, and predict future trends.

5. Green and Sustainable Chemistry: Environmental engineers are exploring innovative green and sustainable chemistry approaches to minimize the use of hazardous substances and promote environmentally friendly processes. This includes developing alternative materials, eco-friendly manufacturing processes, and greener chemical synthesis methods.

6. Nature-Based Solutions: Nature-based solutions involve using natural processes and ecosystems to address environmental challenges. Environmental engineers are working on projects that integrate green infrastructure, such as wetlands, urban forests, and green roofs, to enhance biodiversity, manage stormwater, mitigate urban heat island effects, and improve overall environmental quality.

7. Remote Sensing and Earth Observation: Advances in remote sensing technologies and Earth observation satellites provide valuable data for environmental monitoring and assessment. Environmental engineers utilize these technologies to monitor changes in land cover, detect pollution sources, and assess environmental impacts over large geographic areas.

8. Green Energy and Clean Technologies: The transition to clean and renewable energy sources continues to drive innovations in environmental engineering. Engineers are involved in the development and deployment of technologies such as solar panels, wind turbines, energy storage systems, and electric vehicle infrastructure to accelerate the decarbonization of energy systems.

9. Water Security and Management: Ensuring water

security in the face of growing water scarcity and pollution is a significant challenge. Environmental engineers are involved in developing sustainable water management strategies, including water reuse and recycling, desalination technologies, and efficient irrigation systems.

10. Environmental Justice and Equity: Environmental engineering is increasingly recognizing the importance of addressing environmental justice and equity issues. This involves considering the disproportionate impacts of environmental challenges on marginalized communities and working towards inclusive and equitable environmental solutions.

These emerging trends in environmental engineering reflect the ongoing efforts to address complex environmental challenges and create a more sustainable and resilient future. Environmental engineers play a vital role in developing innovative solutions, leveraging technology, and integrating sustainability principles to tackle these challenges effectively.

Challenges and opportunities in the field

The field of environmental engineering faces various challenges and opportunities as it addresses complex environmental issues. Here are some key challenges and opportunities in the field:

Challenges:

1. Climate Change: Climate change presents a significant challenge, requiring innovative solutions to mitigate greenhouse gas emissions, adapt to changing conditions, and address the impacts on ecosystems and communities.

2. Water Scarcity and Pollution: The growing demand for clean water and the pollution of water sources pose challenges to water resource management. Ensuring access to safe and sufficient water supplies while protecting water ecosystems is a complex task.

3. Sustainable Waste Management: Proper waste management is critical for minimizing environmental impacts and promoting resource efficiency. The challenge lies in developing sustainable waste management systems, including recycling and waste-to-energy technologies, while reducing waste generation.

4. Urbanization and Infrastructure: Rapid urbanization puts pressure on infrastructure systems, including transportation, energy, and waste management. Environmental engineers face the challenge of designing and implementing sustainable and resilient infrastructure to support growing cities.

5. Technological Advancements: While technological

advancements provide opportunities for innovation, they also pose challenges. Environmental engineers must keep pace with emerging technologies and ensure their responsible use to address environmental challenges effectively.

Opportunities:

1. Sustainable Development Goals (SDGs): The SDGs provide a framework and opportunities for environmental engineers to contribute to sustainable development. Addressing the interconnectedness of environmental, social, and economic aspects allows for holistic and integrated solutions.
2. Green and Clean Technologies: The increasing focus on renewable energy, energy efficiency, and clean technologies provides opportunities for environmental engineers to contribute to the transition to a low-carbon economy.
3. Data Analytics and Technology Integration: Advances in data analytics, remote sensing, and modeling techniques offer opportunities to gather and analyze environmental data on a large scale. This data-driven approach can inform decision-making, optimize resource management, and enhance environmental sustainability.
4. Circular Economy and Resource Efficiency: The shift towards a circular economy presents opportunities for environmental engineers to develop innovative solutions for waste reduction, resource recovery, and sustainable production and consumption patterns.
5. Collaboration and Partnerships: Environmental challenges require collaboration among various stakeholders, including governments, industries, communities, and academia. Opportunities lie in forging partnerships to leverage diverse expertise and

resources to address complex environmental issues collectively.

6. Policy and Regulatory Support: Environmental engineers can contribute to the development and implementation of policies and regulations that promote sustainable practices. Engaging with policymakers and advocating for environmentally friendly policies creates opportunities for positive change.

7. Education and Awareness: Opportunities exist to raise awareness and educate the public about environmental issues. Environmental engineers can play a role in informing and empowering individuals and communities to adopt sustainable behaviors and support environmental initiatives.

Navigating these challenges and capitalizing on the opportunities requires interdisciplinary collaboration, innovation, and a systems-thinking approach. Environmental engineers have the opportunity to drive positive change by integrating technical expertise with social, economic, and environmental considerations to create sustainable solutions for a better future.

The role of individuals and organizations in shaping a sustainable future

Individuals and organizations play a crucial role in shaping a sustainable future. Here's a look at their roles:

1. Individuals:
    - Awareness and Education: Individuals can educate themselves about environmental issues and adopt sustainable practices in their daily lives. This includes conserving energy and water, reducing waste, practicing recycling, and choosing eco-friendly products.
    - Consumer Choices: Individuals can make informed consumer choices by supporting environmentally responsible businesses and products. This includes opting for renewable energy, sustainable transportation, and environmentally friendly goods.
    - Advocacy and Activism: Individuals can advocate for sustainable policies and practices by engaging in activism, supporting environmental organizations, and raising awareness within their communities.
    - Sustainable Lifestyles: Individuals can adopt sustainable lifestyles by embracing minimalism, reducing consumption, and embracing alternative modes of transportation such as cycling or public transit.
2. Organizations:

- Sustainability in Operations: Organizations can adopt sustainable practices within their operations. This includes reducing waste, conserving resources, implementing energy-efficient technologies, and prioritizing sustainable supply chain practices.
- Environmental Management Systems: Organizations can implement environmental management systems, such as ISO 14001, to establish policies, set targets, and continuously improve environmental performance.
- Corporate Social Responsibility (CSR): Organizations can incorporate environmental sustainability into their CSR strategies. This includes investing in community initiatives, supporting environmental causes, and engaging in responsible business practices.
- Innovation and Research: Organizations can invest in research and development of sustainable technologies, processes, and products. This includes exploring renewable energy sources, sustainable materials, and eco-friendly alternatives.
- Collaboration and Partnerships: Organizations can collaborate with other businesses, governments, and non-profit organizations to collectively address environmental challenges, share best practices, and drive systemic change.

3. Collaboration between Individuals and Organizations:
- Advocacy and Policy Influence: Individuals and organizations can work together to advocate for sustainable policies at local, national, and international levels. This can involve engaging in public consultations, participating in policy discussions, and supporting environmental

initiatives.

- Knowledge Sharing and Education: Individuals and organizations can collaborate to share knowledge, experiences, and best practices. This can involve hosting workshops, seminars, and conferences to promote sustainability and disseminate information.
- Volunteering and Community Engagement: Individuals and organizations can collaborate on community engagement projects, volunteering efforts, and environmental restoration initiatives to make a positive impact at the local level.
- Funding and Support: Organizations can provide financial support and resources to individuals and grassroots organizations working towards sustainable solutions. This can include grants, sponsorships, and mentorship programs.

By actively participating in sustainable practices and collectively advocating for change, individuals and organizations can contribute to shaping a sustainable future. It is through their combined efforts that meaningful progress can be made in addressing environmental challenges and creating a more sustainable and equitable world.

Recap of key insights and takeaways

Here's a recap of key insights and takeaways:

1. Environmental Engineering: Environmental engineering plays a vital role in addressing complex environmental challenges and promoting sustainable practices. It involves the application of engineering principles, technologies, and management strategies to protect and improve the environment.

2. Sustainability: Sustainability is at the core of environmental engineering. It involves meeting the needs of the present generation without compromising the ability of future generations to meet their own needs. Environmental engineering aims to balance environmental, social, and economic considerations to achieve sustainable outcomes.

3. Environmental Systems: Environmental engineering encompasses various environmental systems, including water, air, land, and waste. Understanding these systems and their interconnections is essential for effective environmental management.

4. Interdisciplinary Approach: Environmental engineering requires an interdisciplinary approach, combining knowledge and expertise from various fields such as engineering, science, policy, and social sciences. Collaboration and integration of diverse perspectives are essential for addressing complex environmental challenges.

5. Sustainable Solutions: Environmental engineering focuses on developing sustainable solutions to

environmental problems. This includes technologies and practices that reduce pollution, conserve resources, promote renewable energy, and mitigate climate change impacts.

6. Environmental Impacts: Environmental engineering projects have both direct and indirect impacts on the environment. Assessing and monitoring these impacts is crucial to ensure that projects are achieving their intended environmental goals and minimizing any adverse effects.

7. Stakeholder Engagement: Engaging stakeholders, including local communities, governments, businesses, and non-profit organizations, is essential for successful environmental engineering projects. Inclusive decision-making processes and active participation of stakeholders foster trust, collaboration, and better project outcomes.

8. Policy and Regulations: Environmental engineering operates within a policy and regulatory framework. Understanding and complying with environmental laws and regulations is crucial for ensuring projects are legally sound and contribute to sustainable development.

9. Technological Advancements: Technological advancements play a significant role in environmental engineering, enabling the development of innovative solutions. Embracing emerging technologies such as data analytics, remote sensing, renewable energy, and smart systems can enhance environmental outcomes.

10. Collaboration and Knowledge Sharing: Collaboration and knowledge sharing among professionals, researchers, communities, and organizations are crucial for advancing environmental engineering. Sharing best practices, lessons learned, and research findings promote continuous learning and improvement.

By incorporating these key insights and takeaways into environmental engineering practices, we can work towards a more sustainable and resilient future, addressing environmental challenges and creating a better world for current and future generations.

Call to action for fostering harmony with nature

Fostering harmony with nature is essential for ensuring a sustainable and thriving future for our planet. Here is a call to action for individuals, organizations, and governments to contribute towards this goal:

1. Embrace Sustainable Practices: As individuals, we can adopt sustainable practices in our daily lives. This includes conserving energy and water, reducing waste, practicing recycling, supporting sustainable products and businesses, and making environmentally conscious choices.

2. Promote Education and Awareness: Education and awareness are key to fostering a deeper understanding of the importance of nature and environmental sustainability. We should support and engage in initiatives that promote environmental education, outreach programs, and campaigns to raise awareness about the interconnectedness of human well-being and the health of the planet.

3. Advocate for Strong Environmental Policies: Individuals and organizations can advocate for the development and implementation of strong environmental policies at local, national, and global levels. This includes supporting policies that promote renewable energy, conservation, sustainable land use, and protection of ecosystems.

4. Support Conservation and Restoration Efforts: Getting

involved in conservation and restoration initiatives can make a significant difference. This can involve volunteering for environmental organizations, participating in local conservation projects, supporting reforestation efforts, and protecting wildlife habitats.

5. Foster Sustainable Business Practices: Businesses have a responsibility to minimize their environmental footprint and integrate sustainability into their operations. Organizations can adopt sustainable practices, reduce emissions, prioritize renewable energy, implement circular economy principles, and engage in responsible supply chain management.

6. Invest in Research and Innovation: Governments, organizations, and individuals can invest in research and innovation for sustainable technologies and solutions. This includes supporting research institutions, funding projects focused on sustainable development, and promoting innovation in areas such as renewable energy, waste management, and ecosystem restoration.

7. Collaborate and Partner: Collaboration among governments, organizations, and communities is crucial for addressing environmental challenges. By working together, we can leverage collective expertise, resources, and knowledge to find innovative solutions and foster greater harmony with nature.

8. Engage in Sustainable Urban Planning: Governments and city planners should prioritize sustainable urban planning practices. This includes designing green spaces, promoting public transportation, implementing energy-efficient buildings, managing waste effectively, and prioritizing the well-being of communities and the environment.

9. Promote Responsible Tourism: As travelers, we can support responsible and sustainable tourism practices. This includes choosing eco-friendly accommodations,

respecting local cultures and environments, minimizing waste generation, and supporting local businesses that prioritize sustainability.

10. Lead by Example: Each individual has the power to inspire and influence others through their actions. By leading by example and embracing sustainable practices, we can inspire others to take action and foster a collective commitment to harmony with nature.

Fostering harmony with nature requires collective effort and a deep sense of responsibility towards our planet and future generations. By taking these actions, we can contribute to a more sustainable and harmonious relationship with nature, ensuring its preservation and the well-being of all life on Earth.

Encouragement for continued exploration and engagement

As we journey towards fostering harmony with nature, it is important to remain engaged and continue exploring sustainable practices. Here is some encouragement for continued exploration and engagement:

1. Embrace Lifelong Learning: The quest for sustainability is an ongoing process that requires continuous learning and growth. Stay curious and seek out opportunities to expand your knowledge about environmental issues, emerging technologies, and innovative solutions. Explore books, articles, documentaries, and online resources to deepen your understanding.

2. Engage in Dialogue: Engage in meaningful conversations with individuals from diverse backgrounds and perspectives. Discuss ideas, share experiences, and learn from each other's insights. Engaging in dialogue fosters collaboration, broadens perspectives, and inspires innovative approaches to environmental challenges.

3. Be an Advocate: Use your voice to advocate for environmental causes. Share your knowledge, experiences, and passion with others. Write letters to policymakers, participate in public consultations, or join environmental organizations to contribute to collective efforts in creating positive change.

4. Support Local Initiatives: Get involved in local environmental initiatives and grassroots movements.

Support community-led projects that aim to protect and restore natural habitats, promote sustainable practices, and engage in conservation efforts. By supporting local initiatives, you can make a tangible impact in your own community.

5. Collaborate and Network: Seek opportunities to collaborate with like-minded individuals, organizations, and communities. Engage in partnerships and networks that focus on environmental sustainability. Collaborative efforts amplify individual actions and create a collective force for change.

6. Celebrate Successes: Celebrate and acknowledge the successes and achievements in sustainability. Recognize individuals, organizations, and communities that are making a positive difference. By highlighting and celebrating successes, we inspire and motivate others to take action.

7. Stay Resilient: The journey towards a sustainable future may encounter obstacles and setbacks. It is important to stay resilient and persevere in the face of challenges. Learn from failures, adapt strategies, and remain committed to the goal of fostering harmony with nature.

8. Lead by Example: Be a role model by incorporating sustainable practices into your own life. Demonstrate how small actions can have a ripple effect and inspire others to follow suit. Lead by example and encourage others to join you in making a positive impact.

9. Practice Gratitude: Cultivate a sense of gratitude for the beauty and gifts of the natural world. Appreciate the wonders of nature and the interconnectedness of all living beings. By nurturing gratitude, we deepen our commitment to protecting and preserving the environment.

10. Inspire Others: Share your passion, knowledge, and experiences with others. Inspire and empower

individuals to take action and make sustainable choices. Be a catalyst for change by influencing those around you to embark on their own journey towards fostering harmony with nature.

Remember that every action, no matter how small, contributes to the collective effort of creating a sustainable future. Stay engaged, continue exploring, and never underestimate the power of your individual and collective actions in making a positive impact on the world.